HACKING YOUR LEGO® MINDSTORMS® EV3 KIT

John Baichtal

800 East 96th Street,
Indianapolis, Indiana 46240 USA

Hacking Your LEGO® Mindstorms® EV3 Kit

ISBN-13: 978-0-7897-5538-4
ISBN-10: 0-7897-5538-6

Library of Congress Control Number: 2015942445

Printed in the United States of America

First Printing: November 2015

Trademarks

All terms mentioned in this book that are known to be trademarks or service marks have been appropriately capitalized. Que Publishing cannot attest to the accuracy of this information. Use of a term in this book should not be regarded as affecting the validity of any trademark or service mark. LEGO® and MINDSTORMS® are registered trademarks of The LEGO Group. This book is not authorized or endorsed by the LEGO® Group.

Warning and Disclaimer

Every effort has been made to make this book as complete and as accurate as possible, but no warranty or fitness is implied. The information provided is on an "as is" basis. The author and the publisher shall have neither liability nor responsibility to any person or entity with respect to any loss or damages arising from the information contained in this book.

Special Sales

For information about buying this title in bulk quantities, or for special sales opportunities (which may include electronic versions; custom cover designs; and content particular to your business, training goals, marketing focus, or branding interests), please contact our corporate sales department at corpsales@pearsoned.com or (800) 382-3419.

For government sales inquiries, please contact governmentsales@pearsoned.com.

For questions about sales outside the U.S., please contact international@pearsoned.com.

Editor-in-Chief
Greg Wiegand

Executive Editor
Rick Kughen

Development Editor
Susan Arendt

Managing Editor
Sandra Schroeder

Project Editor
Mandie Frank

Copy Editor
Geneil Breeze

Senior Indexer
Cheryl Lenser

Proofreader
Jess DeGabriele

Technical Editor
James Floyd Kelly

Editorial Assistant
Kristen Watterson

Designer
Mark Shirar

Compositor
Studio GaLou

Contents at a Glance

Table of Contents

About the Author

John Baichtal has written or edited more than a dozen books, including the award-winning *Cult of Lego* (2011, No Starch Press), *Make: LEGO and Arduino Projects* (2012, Maker Media) with Adam Wolf and Matthew Beckler, *Robot Builder* (2014, Que), and *Basic Robot Building with LEGO Mindstorms NXT 2.0* (2012, Que), as well as *Building Your Own Drones* (2015, Que). His most recent book is *Maker Pro* (2014, Maker Media), a collection of essays and interviews describing life as a professional maker. John lives in Minneapolis, Minnesota with his wife and three children.

Dedication

This book is dedicated to my wife Elise, my kids Arden, Rosemary, and Jack, my mom Barbara, and to all those who strive to make an item or platform work better for them by hacking it!

Acknowledgments

Thanks for the inspiration and assistance (in no particular order) to Miguel Valenzuela, Pete McKenna, Steve Norris, Steven Anderson, MakerBeam, Jude Dornisch, SparkFun Engineering, Adam Wolf, Michael Freiert, Sophi Kravitz, Christina Zhang, Lenore Edman, Rick Kughen, Sean Michael Ragan, John Wilson, Susan Solarz, Akiba, Mark Frauenfelder, Chris Berger, Michael Krumpus, Alex Dyba, Brian Jepson, Becca Steffen, Dave Bryan, Actobotics, Mike Hord, Makeblock, Pat Arneson, Erin Kennedy, Mindsensors, Windell H. Oskay, Johngineer, Matthew Beckler, Riley Harrison, David Lang, Trammell Hudson, Kristina Durivage, AnnMarie Thomas, Pete Prodoehl, Bruce Shapiro, Alex Allmont, John Edgar Park, and Dexter Industries. Apologies to anyone I forgot!

We Want to Hear from You!

As the reader of this book, *you* are our most important critic and commentator. We value your opinion and want to know what we're doing right, what we could do better, what areas you'd like to see us publish in, and any other words of wisdom you're willing to pass our way.

We welcome your comments. You can email or write to let us know what you did or didn't like about this book—as well as what we can do to make our books better.

Please note that we cannot help you with technical problems related to the topic of this book.

When you write, please be sure to include this book's title and author as well as your name and email address. We will carefully review your comments and share them with the author and editors who worked on the book.

Email: feedback@quepublishing.com

Mail: Que Publishing
ATTN: Reader Feedback
800 East 96th Street
Indianapolis, IN 46240 USA

Reader Services

Visit our website and register this book at quepublishing.com/register for convenient access to any updates, downloads, or errata that might be available for this book.

Introduction

Building a robot teaches you a lot about engineering, electronics, and mechanics. That's what the LEGO Group had in mind when it began developing its signature robotics set, Mindstorms, back in 1998.

The latest edition, LEGO Mindstorms EV3, represents the culmination of what the company learned from several unsuccessful robot sets. (Do Cybermaster or Spybotics sound familiar? Didn't think so.) as well as the two previous editions of the Mindstorms product. Like its predecessors, EV3 consists of a plastic building set with compatible motors, sensors, and a microcontroller brick that runs everything (see Figure 1.1).

Mindstorms has come to be thought of as a stepping stone to "real" robotics because all the principles are the same. This so-called toy teaches you how to program a microcontroller, build gear assemblies, and create reinforced structures. If you want to learn about engineering principles, you can do a lot worse than Mindstorms.

FIGURE 1.1 The EV3 Intelligent Brick follows a program to control your robot.

Hacking Mindstorms

A robot-building set offers many possibilities, but there's always room for more options. Sooner or later you get the idea you want to change something about the set. Maybe you need the perfect beam to complete your robot, but it's a part not found in the Mindstorms inventory. Possibly you want to add a third-party sensor to your robot. Seems reasonable, right?

Well, not to everyone. Quite honestly, some perfectly sensible people are opposed to hacking the Mindstorms set in any way. The challenge for them is in accomplishing their objectives within the limitations of the set. They refuse to include any non-LEGO parts. They do not glue the pieces, nor cut or drill them.

Although I admire the tenacity and inventiveness of these stalwarts, I choose a different path: I just hack it. If the microcontroller doesn't work the way I want it to, I reprogram it or use an Arduino instead, using the Bricktronics shield, shown in Figure 1.2, to control standard Mindstorms sensors and motors.

FIGURE 1.2 The Bricktronics shield interfaces LEGO's proprietary motors and sensors with an Arduino.

I can add a third-party electronic module such as Dexter Industries' dCompass to my model (see Figure 1.3), giving it the capability to tell what direction is north. This is just one example of dozens of cool products that you can purchase to accessorize your EV3 set.

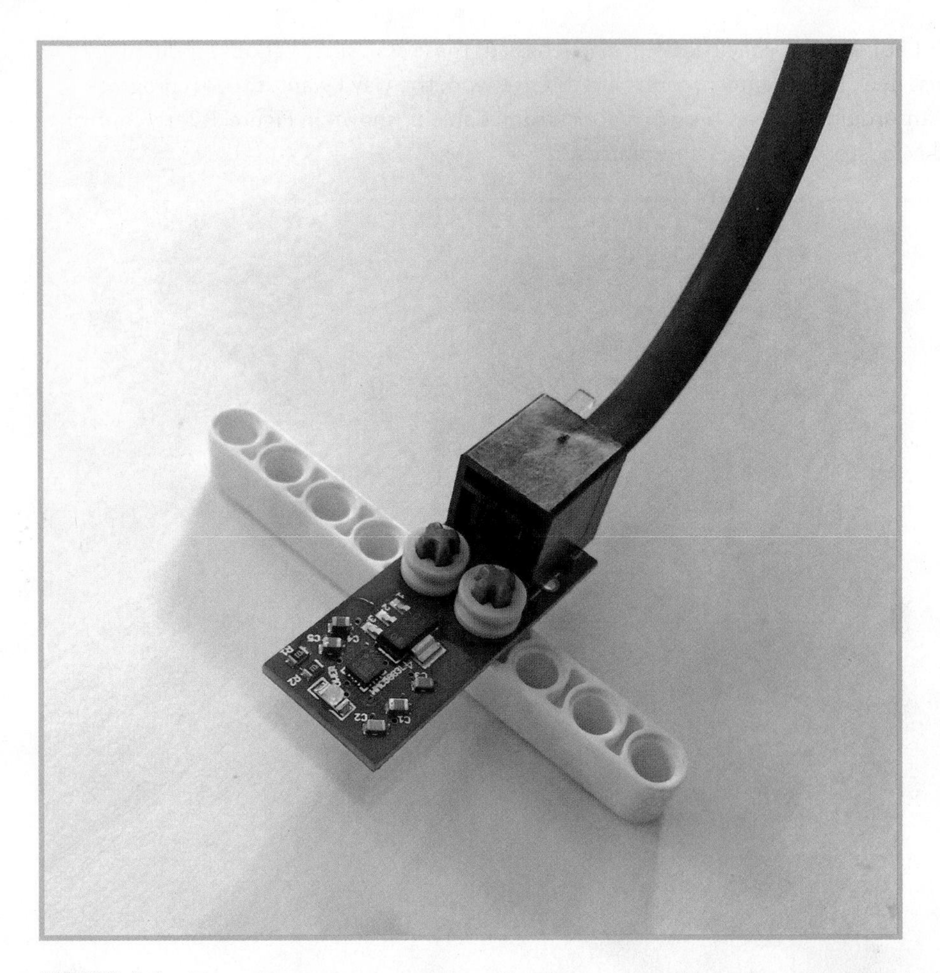

FIGURE 1.3 The dCompass module tells the robot which direction it's facing.

If a part isn't working for me, I modify it or replace it with another part that does work—even if I have to create the new one myself. The hole-studded boards in Figure 1.4 match up with LEGO hole spacing, enabling me to reinforce models or even make box-shaped enclosures.

Makeblock, another building set with Mindstorms-compatible hole spacing, not only gives you more parts to work with but adds much-needed strength and rigidity as its beams extrude out of aluminum.

Hacking your LEGO Mindstorms EV3 set has never been easier!

FIGURE 1.4 The laser-cut holes in these wooden shapes match up with LEGO beams.

Chapter Topics

I just touched on a few of the possibilities. You learn many more techniques for making the most of your Mindstorms set:

- In Chapter 2, “Project: Plotter Bot,” you build the first project of the book, a drawing robot that decorates a 4x6 index card with a pen.
- It’s back to school in Chapter 3, “Hacking LEGO I: Connections,” where you learn about Mindstorms’ wires and how to hack them. The chapter also explores a bevy of wireless control options.
- This know-how is put to good use in Chapter 4, “Project: Remote-Controlled Crane,” where you build a mobile crane that rolls along a railing and lowers a hook.
- Chapter 5, “Hacking LEGO II: Alternate Microcontrollers,” teaches you how to swap in Arduino microcontrollers and Raspberry Pi and BeagleBone Black microcomputers in place of the trusty but dull Mindstorms Intelligent Brick that ships with the set.
- Chapter 6, “Project: Robot Flower,” shows you how to build a robotic flower using the Mindstorms Intelligent Brick as well as an Arduino microcontroller.
- Chapter 7, “Hacking LEGO III: Create Your Own LEGO Parts,” shows you how to 3D-print, laser, and mill your own LEGO-compatible beams when you can’t find that “perfect” part.
- Chapter 8, “Project: Ball Contraption,” shows you how to create an intriguing device whose sole purpose is to route a LEGO ball around a series of chutes and ramps. You also have an opportunity to create a few simple LEGO-compatible parts to add on to your contraption.
- In Chapter 9, “Hacking LEGO IV: Add-on Electronics,” you explore the variety of third-party sensors that you can control with your Mindstorms Intelligent Brick.
- Chapter 10, “Project: Flagpole Climber,” shows you how to make a pole-climbing robot. You also learn how to hook up an altimeter so you know how far up it goes.
- The book concludes with a concise glossary to clarify any of those cryptic Mindstorms and robotics terms.

This book is dedicated to exploring just a few of the amazing hacks out in the wild. Make robots, have fun, and don’t worry about building something perfect.

2

Project: Plotter Bot

The first project creates a drawing robot that resembles the device known as the plotter—essentially a robot that manipulates motors to draw with a pen (see Figure 2.1). Professional plotters have precisely turning motors whose rotations are divided into hundreds of steps (they're called *stepper motors*, unsurprisingly), which lets them draw shapes with great precision. This robot is not precise, but it is fun, and you get to build and program it yourself with nearly everything found in your EV3 set.

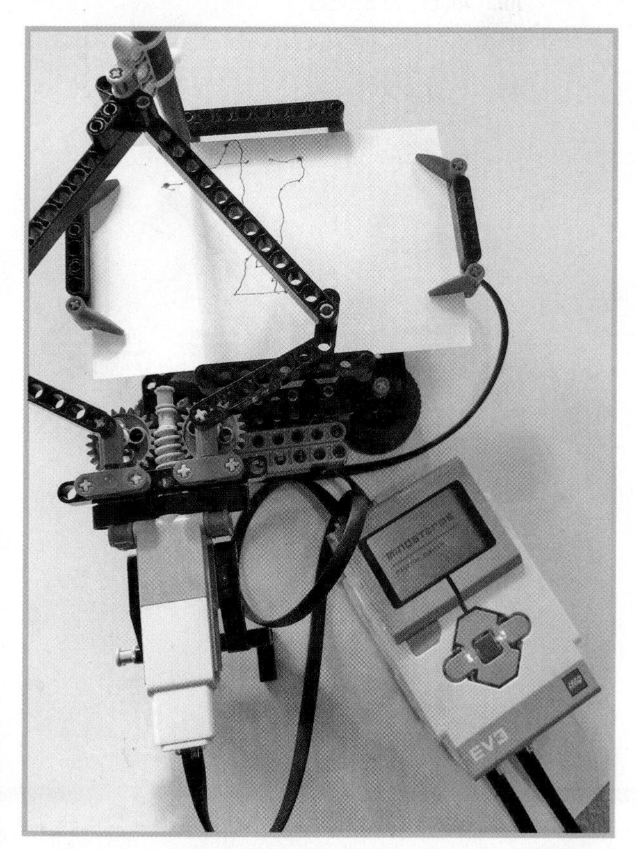

FIGURE 2.1 The PPlotter Bot draws on an index card.

Parts List

Let's get started! You need the following parts to build your plotter bot; you should be able to find all of them in a fully stocked EV3 kit. You can see the parts in Figure 2.2.

A. 1x EV3 Intelligent Brick

B. 1x large motors

C. 1x medium motor

D. 2x 15M beams

E. 4x 13M beams

F. 3x 11M beams

G. 6x 9M beams

H. 5x 7M beams

I. 5x 5M beams

J. 7x 3M beams

K. 2x 3x7 angle beams

L. 4x 3x5 90-degree angle beams

M. 2x 4x4 angle beams

N. 6x 3x7 double-angle beams

O. 1x T-beam

P. 2x4 90-degree angle beams

Q. 2x 5x11 frame bricks

R. 2x 5x7 frame bricks

S. 2x 8M cross axle with end stop

T. 1x 7M cross axle

U. 1x 6M cross axle

V. 7x 5M cross axles

W. 5x 3M cross axles

X. 2x 2M cross axles

Y. 2x small tires and rims

Z. 2x wedge-belt tires and rims

AA. 4x normal tires and rims

BB. 1x 3X2 cross block

CC. 8x 3M beam with pegs

DD. 10x double cross blocks

EE. 1x worm drive

FF. 2x 0-degree angle element

GG. 2x Z24 gears

HH. 4x cross blocks

II. 4x Bionicle eyes

JJ. Pegs and bushings: You need all kinds of pegs and both regular and half bushings.

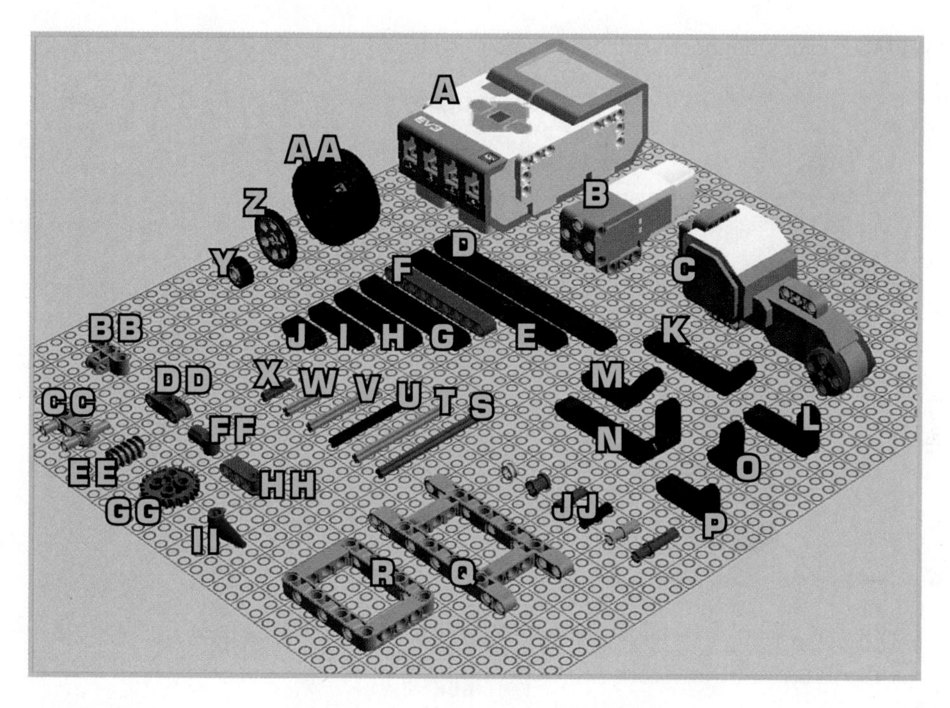

FIGURE 2.2 You'll need these Mindstorms parts to build the Pplotter BBot.

In addition, you need the following non-LEGO parts that you can find nearly anywhere:

- Small zip ties
- 4x6 index cards
- Felt-tipped pen

Building the Plotter Bot

One of the cool things about LEGO (including Mindstorms) is that you don't need any tools or special hardware—you just grab parts and start building. In that spirit, let's begin building the Plotter Bot. Note that I use blue-colored bricks to mark new parts for each step; otherwise, the colors depicted conform to the parts found in the EV3 set.

STEP 1 Add four cross connectors to the motor as shown in Figure 2.3.

FIGURE 2.3

STEP 2 Connect a pair of 2x4 angle beams to the motor connectors (see Figure 2.4).

FIGURE 2.4

STEP 3 Add two light gray connector pegs—these don't have the friction tabs, so elements rotate freely on them. Figure 2.5 shows how it should look.

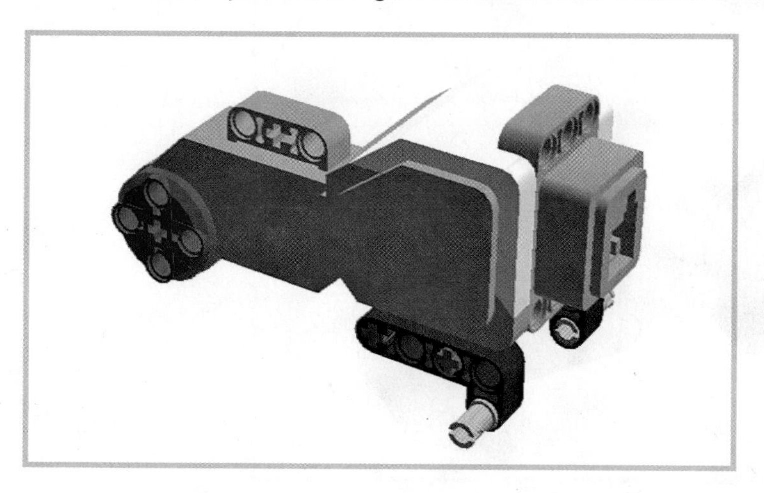

FIGURE 2.5

STEP 4. Put the little wheels on the hubs and attach them to the gray connectors, as shown in Figure 2.6.

FIGURE 2.6

STEP 5 Insert a 7M axle into the motor's hub, centering it like you see in Figure 2.7.

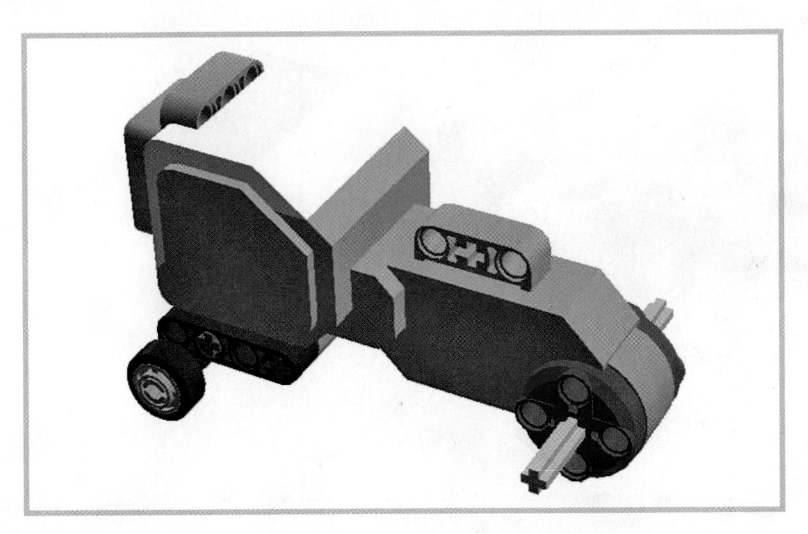

FIGURE 2.7

STEP 6 Next, secure the axle with a pair of bushings. Figure 2.8 shows how it should look—the other bushing is hidden behind the model.

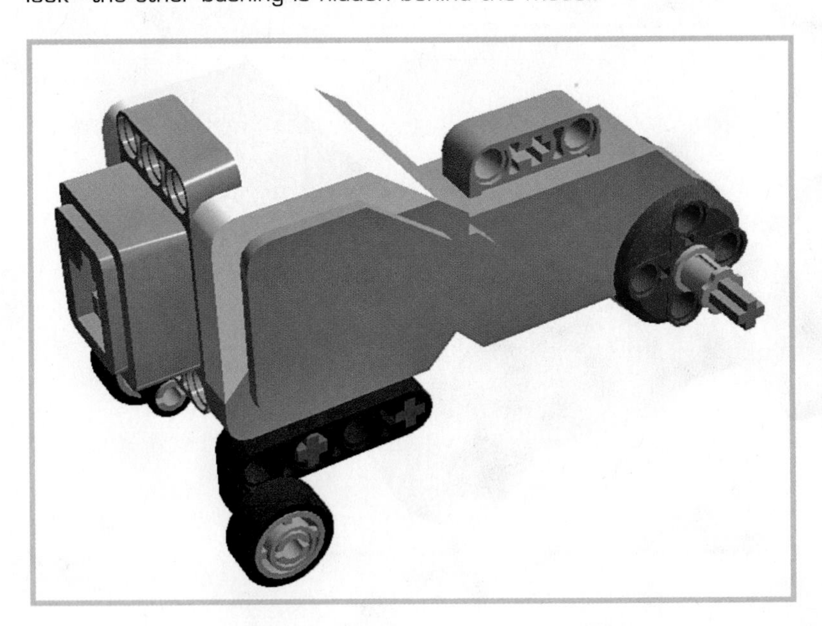

FIGURE 2.8

STEP 7 Put a regular connector peg in the motor as shown in Figure 2.9.

FIGURE 2.9

STEP 8 Add the wheels, as shown in Figure 2.10. These look sweet, but I wish they had better traction.

FIGURE 2.10

STEP 9 Slide two axles, a 3M and a 5M, through the motor's mounting holes. You can see how it should look in Figure 2.11. (The M refers to the length; a 5M axle is the same length as a 5-hole beam.)

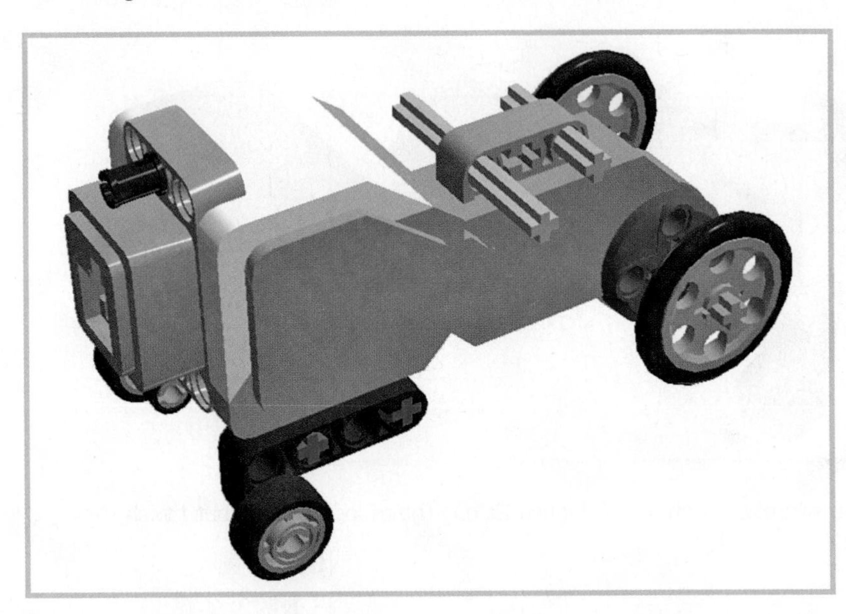

FIGURE 2.11

STEP 10 Three cross blocks get added next, two to the axles and one to the connector on the back of the motor (see Figure 2.12).

FIGURE 2.12

STEP 11 Insert two 3M connector pegs into the holes on the cross blocks you just added. Figure 2.13 shows how it should look.

FIGURE 2.13

STEP 12 Attach a pair of 5M beams as shown in Figure 2.14.

FIGURE 2.14

STEP 13 A pair of angle beams serve as bumpers to help the carriage roll back and forth more easily (see Figure 2.15).

FIGURE 2.15

STEP 14 A slew of pegs are added: two cross connectors onto the cross block and three regular black connector pegs into the 5M beams. Figure 2.16 shows how it should look.

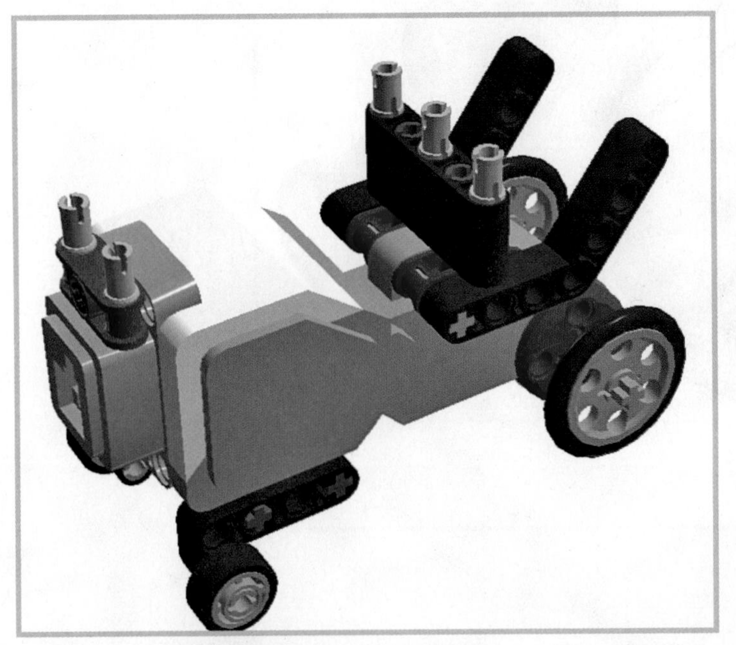

FIGURE 2.16

STEP 15 Two 13M beams support the print bed, as shown in Figure 2.17.

FIGURE 2.17

STEP 16 Let's work on the actual bed. Take a 5x11 chassis brick and insert five 3M connector pegs as shown in Figure 2.18. A 9M beam fits over the pegs.

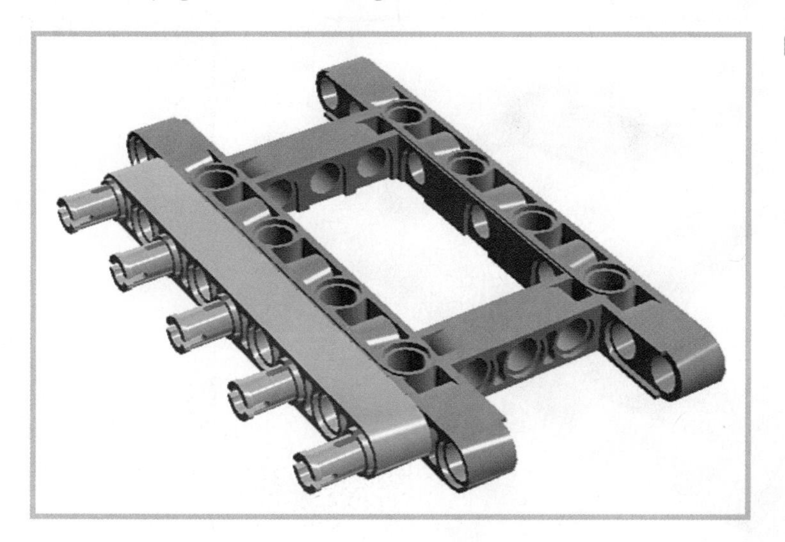

FIGURE 2.18

STEP 17 Add another chassis brick, as shown in Figure 2.19.

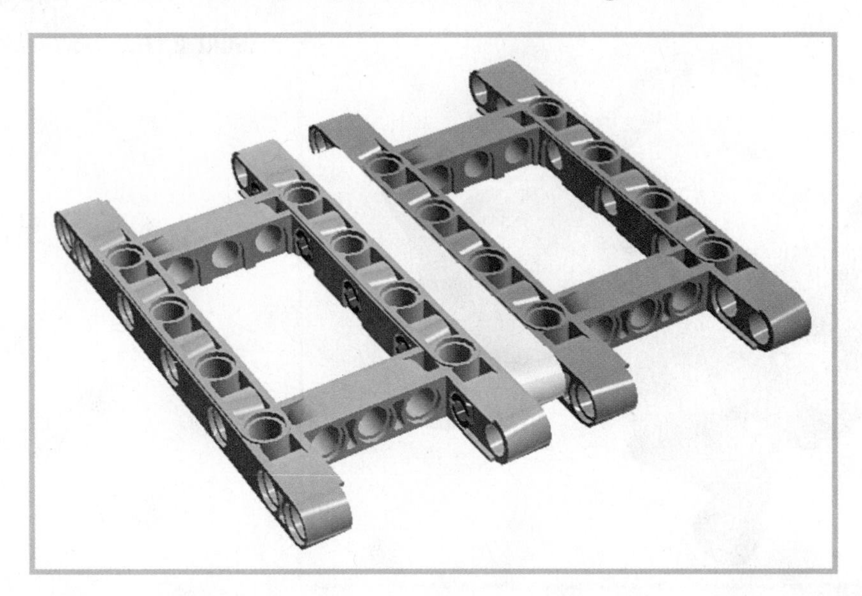

FIGURE 2.19

STEP 18 Insert four cross connector pegs into the chassis bricks' mounting holes (see Figure 2.20).

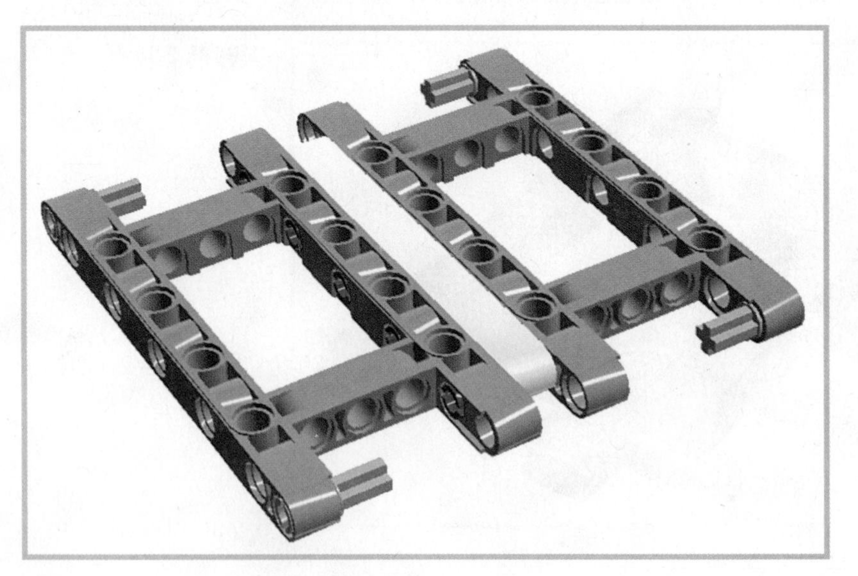

FIGURE 2.20

STEP 19 Four cross blocks are attached to the model, as shown in Figure 2.21.

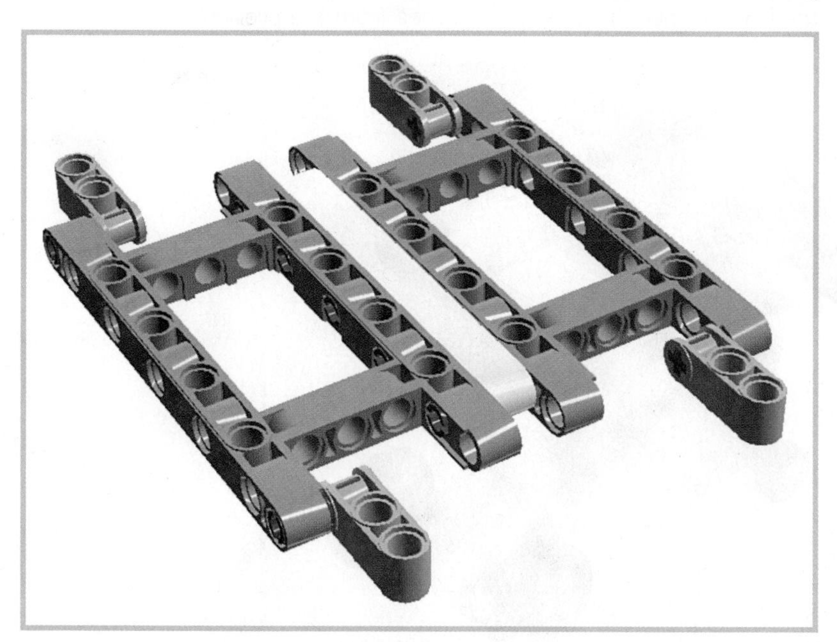

FIGURE 2.21

STEP 20 Add four black connector pegs, as shown in Figure 2.22. These connect the print bed to the carriage.

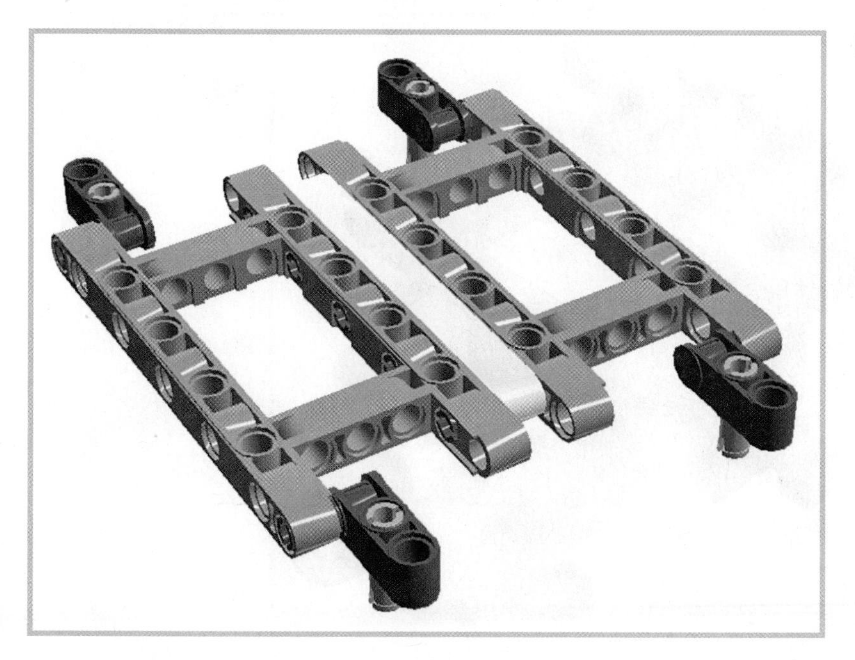

FIGURE 2.22

STEP 21 Attach the bed to the carriage using the pegs you just added (see Figure 2.23). Note that the bed has been rotated 90 degrees from the previous step.

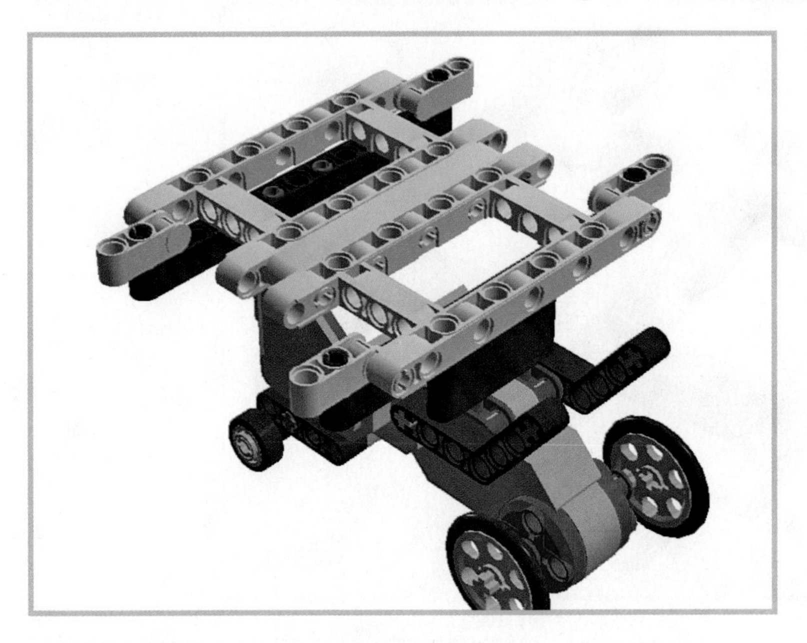

FIGURE 2.23

STEP 22 Add four connector pegs as shown in Figure 2.24.

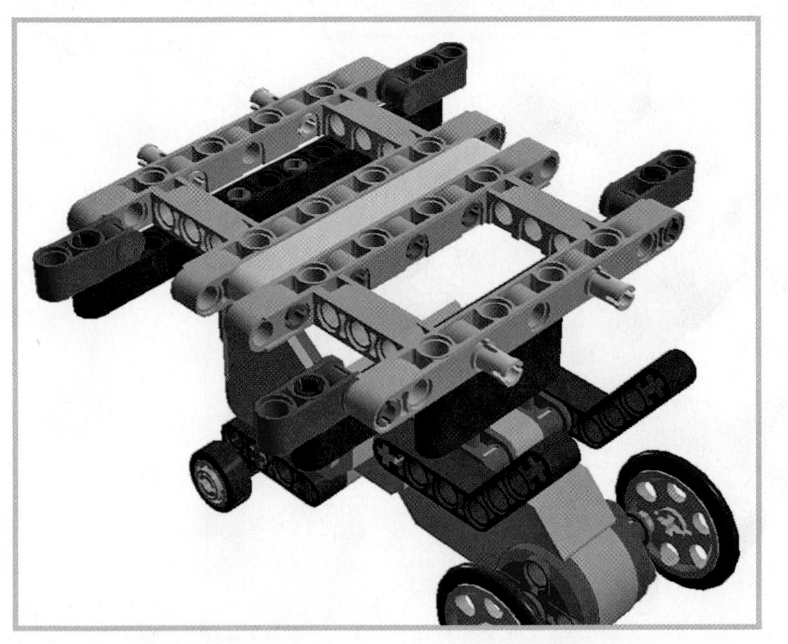

FIGURE 2.24

STEP 23 Two 5x7 chassis frames are added next, as shown in Figure 2.25.

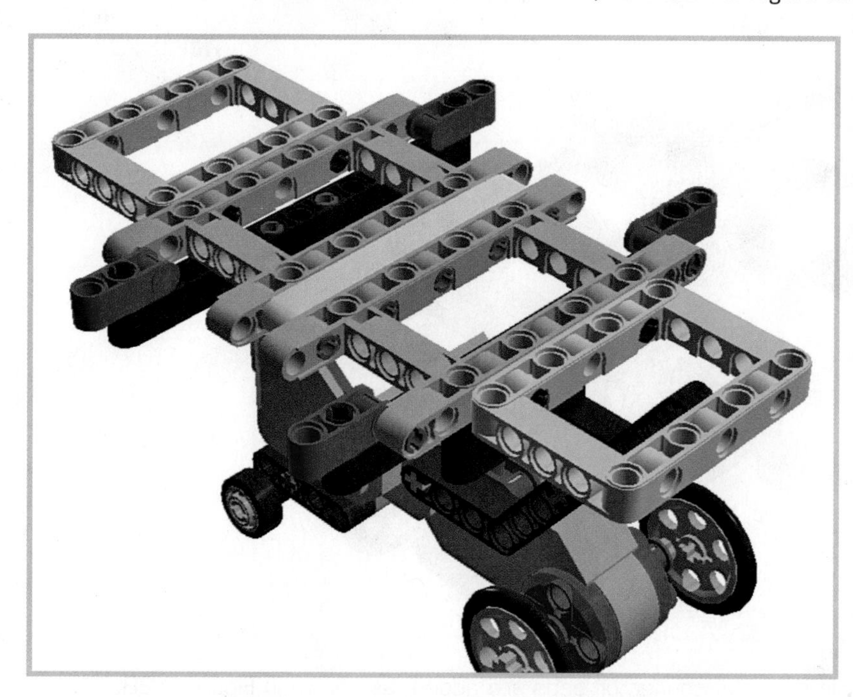

FIGURE 2.25

STEP 24 Four 3M beams with pegs are added to the chassis bricks (see Figure 2.26).

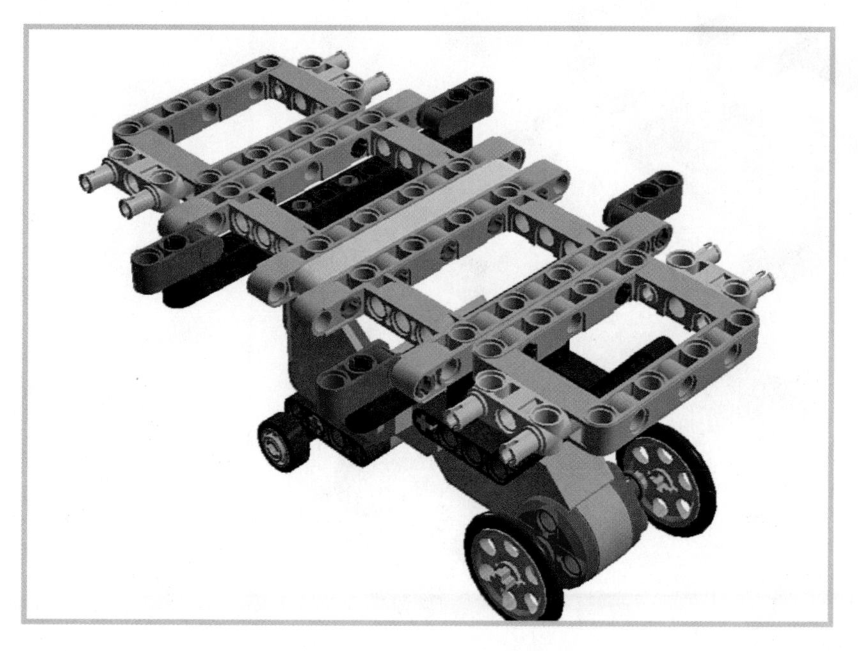

FIGURE 2.26

STEP 25 3M beams are added next. It should look like Figure 2.27.

FIGURE 2.27

STEP 26 A whole bunch of connector pegs are added, as shown in Figure 2.28.

FIGURE 2.28

STEP 27 Next, add the railings to help keep the index card from moving. 9M beams are added to the sides and 5M beams to the ends. Figure 2.29 shows how it should look.

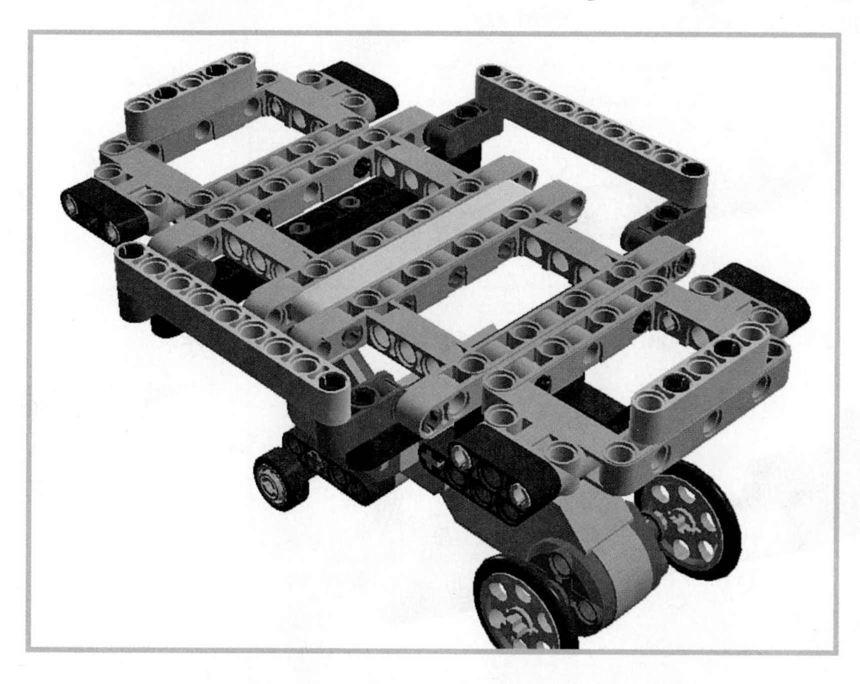

FIGURE 2.29

STEP 28 To complete the carriage, add Bionicle eyes and cross connectors (see Figure 2.30). You can rotate these parts to help keep the index card in place.

FIGURE 2.30

STEP 29 Next, let's begin work on the robot's base that includes the motor controlling the X axis. It begins with a red 11M beam with two connector pegs and two 3M pegs added to it. Figure 2.31 shows how it should look.

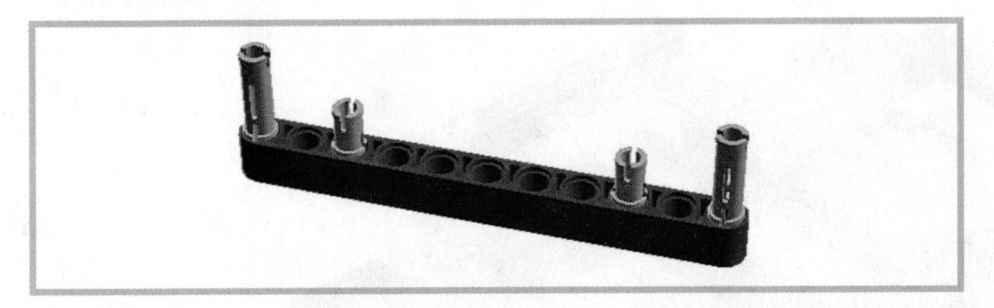

FIGURE 2.31

STEP 30 Add a pair of 3x5 angle beams to the assembly (see Figure 2.32).

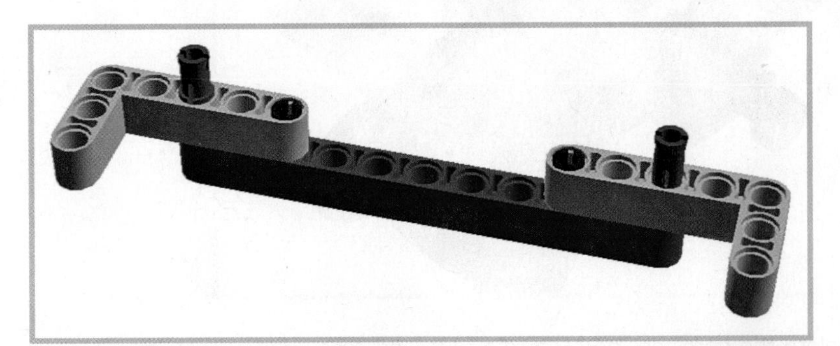

FIGURE 2.32

STEP 31 Continue building the base with the two 15M beams pictured in Figure 2.33. Note the two connector pegs and two cross connectors are added as well.

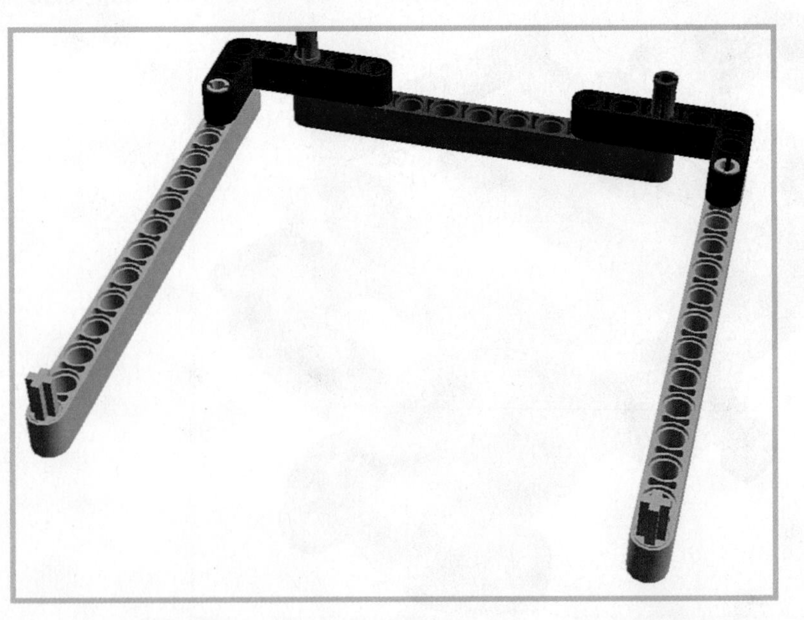

FIGURE 2.33

STEP 32 Add the double-angle beams pictured in Figure 2.34.

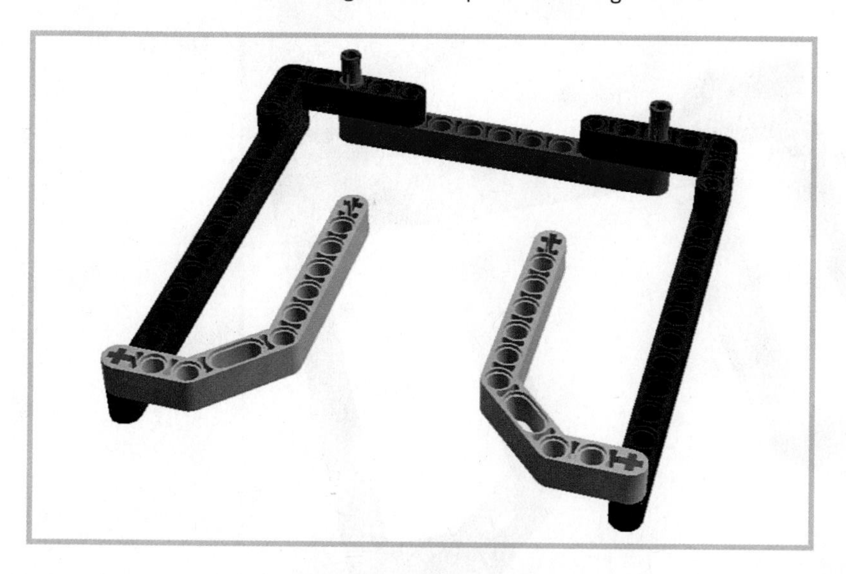

FIGURE 2.34

STEP 33 Connect a bunch of pegs, both the regular kind as well as cross connectors. Figure 2.35 shows where they should go.

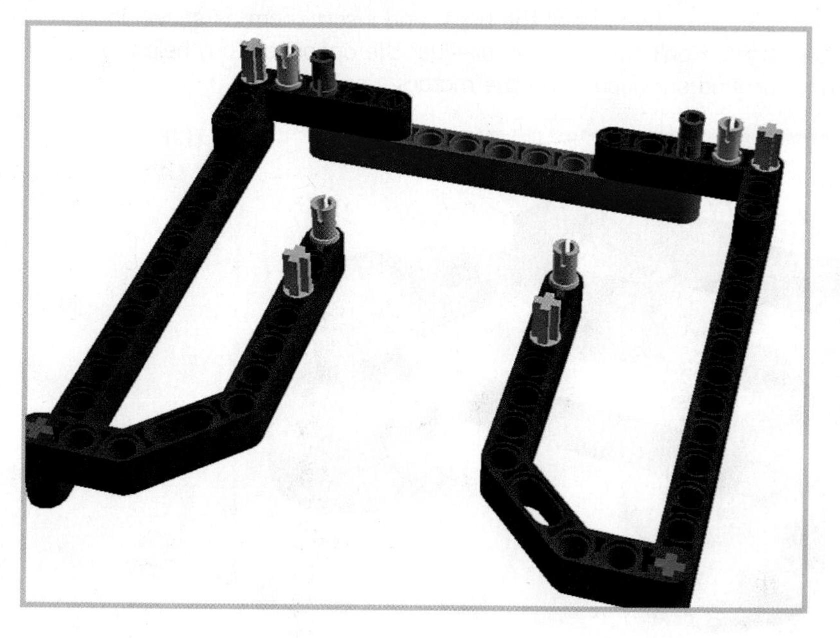

FIGURE 2.35

STEP 34 Two more double-angle beams reinforce the base (see Figure 2.36).

FIGURE 2.36

STEP 35 Insert 6M axles into the hubs of the tires, and insert them as shown in Figure 2.37. Those wheels won't move the robot—just the opposite! They help the Plotter Bot not move around so much when the motors work.

FIGURE 2.37

STEP 36 Secure the feet with bushings, as shown in Figure 2.38.

FIGURE 2.38

STEP 37 Add four 3M beams with pegs, as shown in Figure 2.39. This is the beginning of the platform that will hold the X-axis motor.

FIGURE 2.39

STEP 38 Next, add four connector pegs and a 7M beam (see Figure 2.40).

FIGURE 2.40

STEP 39 Attach two 9M beams and a 7M beam as shown in Figure 2.41. Six connector pegs are added as well. Note that the 7M beam doesn't connect to the beam it's resting on—just hold it in place for now!

FIGURE 2.41

STEP 40 Let's add a pair of 5M beams and four connector pegs, as shown in Figure 2.42. Note that the assembly has turned 180 degrees from the previous step.

FIGURE 2.42

STEP 41 Add two 7M beams and six more pegs. Figure 2.43 shows how it should look.

FIGURE 2.43

STEP 42 A pair of red 11M beams and four pegs are next (see Figure 2.44).

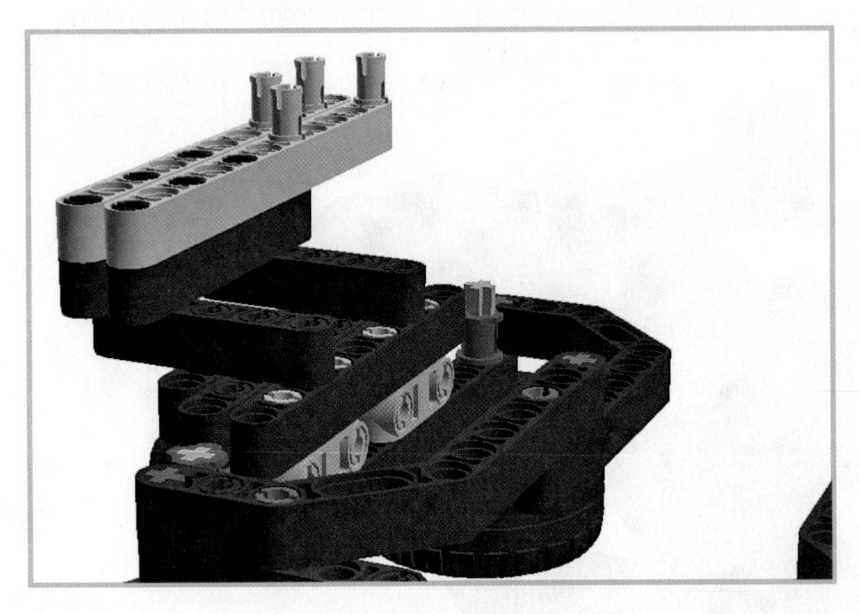

FIGURE 2.44

STEP 43 Add a pair of 2x4 angle beams along with a pair of red 2M axles and 2 connector pegs, as shown in Figure 2.45.

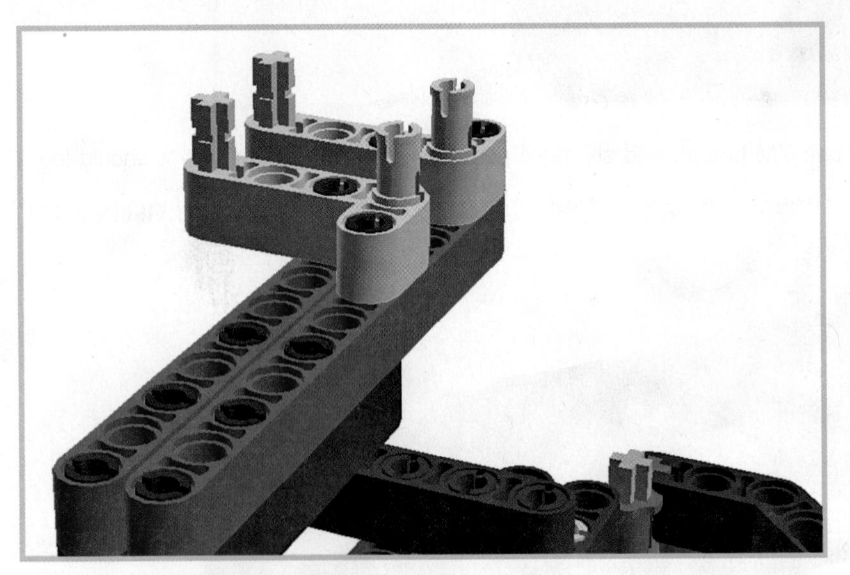

FIGURE 2.45

STEP 44 Attach a 5M beam and a cross block to the assembly. Figure 2.46 shows how it should look.

FIGURE 2.46

STEP 45 The motor comes next, as shown in Figure 2.47. Connect it to the motor mount with a single connector peg. This seems precarious, but trust me, it will be reinforced!

FIGURE 2.47

STEP 46 Attach two pegs and two cross connectors to the motor, one of each on a side. Figure 2.48 shows how it should look.

FIGURE 2.48

STEP 47 A pair of double-angle beams give the motor some much-needed support. Note the extra connector pegs at the bottom. Figure 2.49 shows how it should look.

FIGURE 2.49

STEP 48 A pair of 3x5 angle beams serve as the motor's legs (see Figure 2.50).

FIGURE 2.50

STEP 49 Use a pair of 8M axles with end stops ("S" in the Parts List) to trap three 3M beams in between the double-angle beams, as shown in Figure 2.51.

FIGURE 2.51

STEP 50 Add four cross connectors (not visible in Figure 2.52) and two cross blocks to the motor, as shown in Figure 2.52.

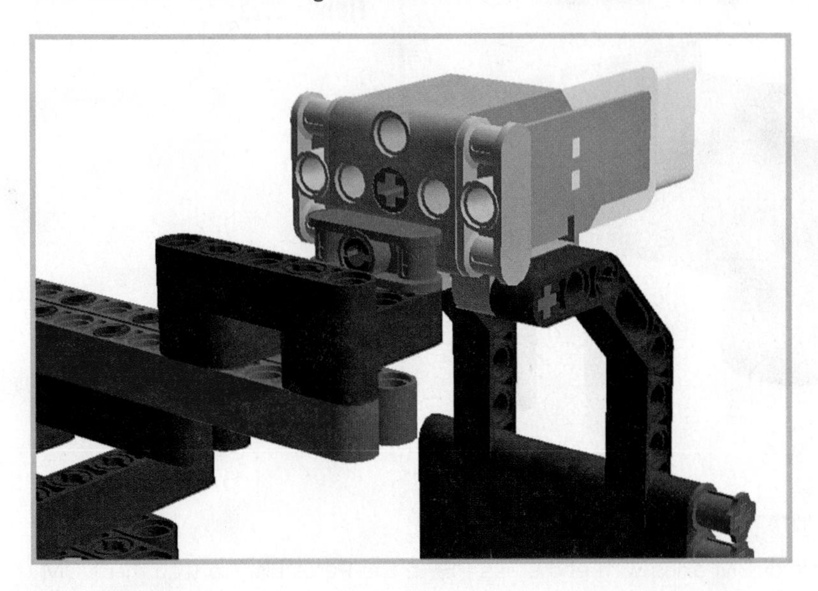

FIGURE 2.52

STEP 51 Add a 6M axle to the motor's hub and a pair of 3M pegs to either side (see Figure 2.53).

FIGURE 2.53

STEP 52 A 7M beam is placed across the motor's hub, as shown in Figure 2.54.

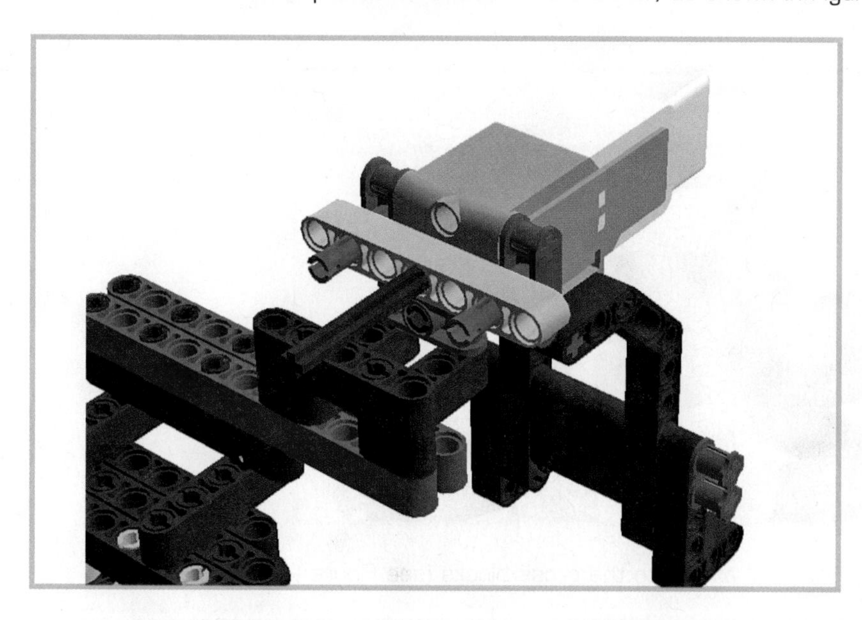

FIGURE 2.54

STEP 53 Add a bushing, the worm drive, and a half-bushing to the axle. Figure 2.55 shows how it should look.

FIGURE 2.55

STEP 54 Attach a pair of cross blocks to the ends of the 3M pegs, as shown in Figure 2.56.

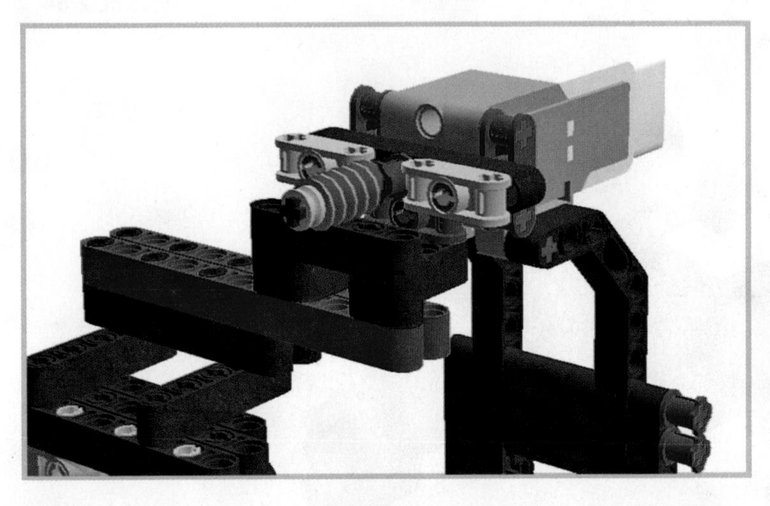

FIGURE 2.56

STEP 55 Insert four 3M axles into the cross blocks (see Figure 2.57).

FIGURE 2.57

STEP 56 Lay a 9M beam along the top of the assembly, threading the 3M axles through the holes in the beam, as shown in Figure 2.58.

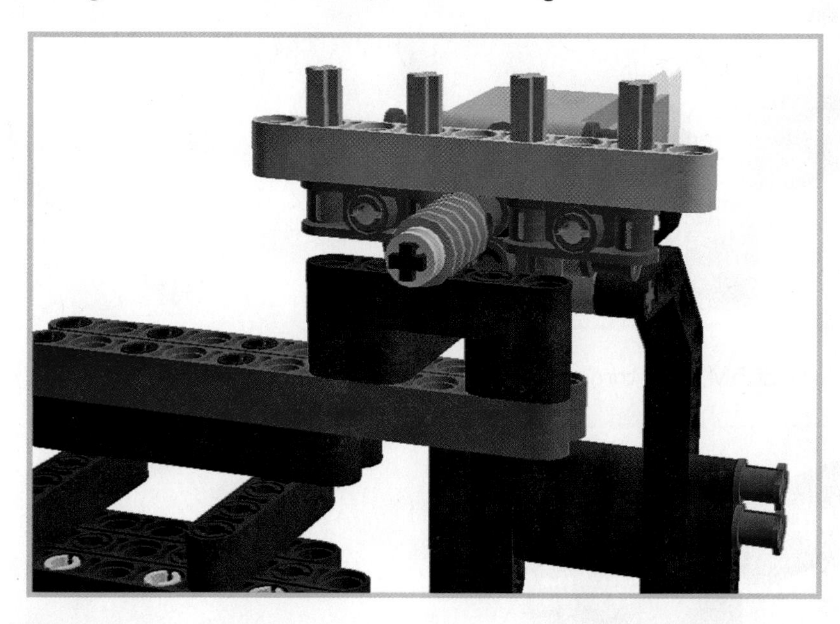

FIGURE 2.58

STEP 57 Let's work on the gear assemblies. Take a pair of cross blocks and add cross connectors (see Figure 2.59).

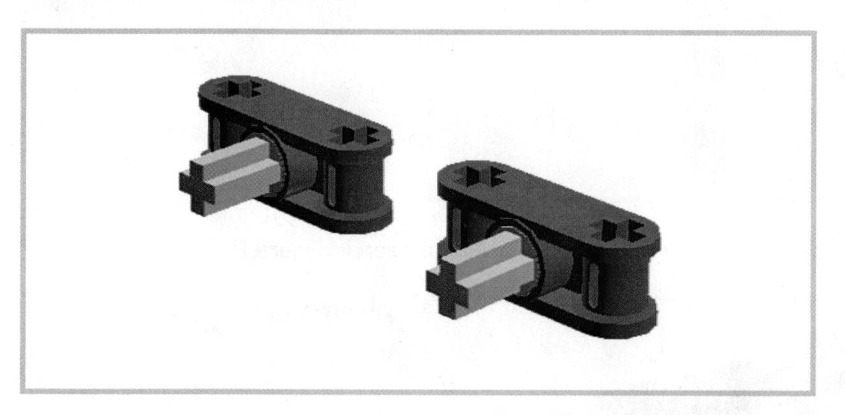

FIGURE 2.59

STEP 58 Add 0-degree angle elements (which is a very silly name) to the cross connectors. Figure 2.60 shows how it should look.

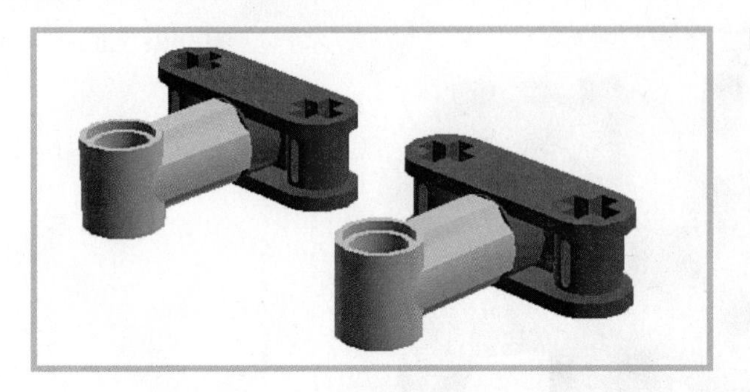

FIGURE 2.60

STEP 59 Slide a pair of 5M axles through the angle elements, as shown in Figure 2.61.

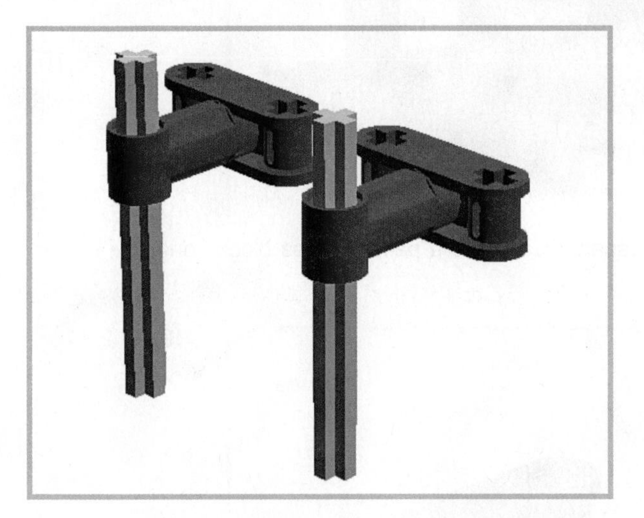

FIGURE 2.61

STEP 60 Reinforce the axles with a pair of bushings, seen in Figure 2.62.

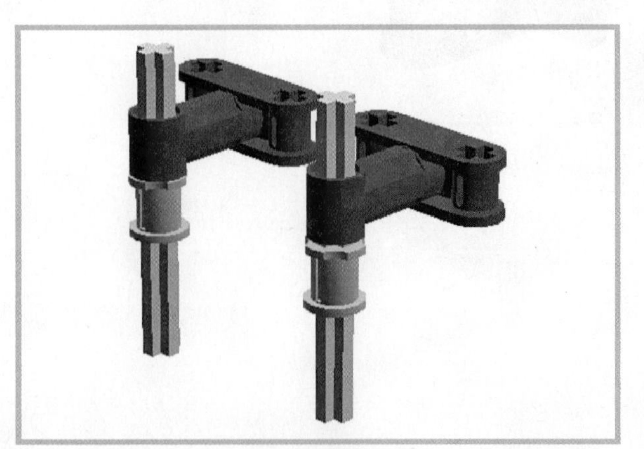

FIGURE 2.62

STEP 61 Slide a pair of 24-tooth gears onto the axles as shown in Figure 2.63. These drive the X-axis arms as they carry the pen.

FIGURE 2.63

STEP 62 Install the gear assemblies onto the top of the motor, sliding the cross axles onto the exposed 3M axles. Figure 2.64 shows how it should look.

FIGURE 2.64

STEP 63 Add 3M angle beams to the tops of the axles. Also add a pair of no-friction (gray) connector pegs as seen in Figure 2.65.

FIGURE 2.65

STEP 64 Connect a pair of 13M beams to the gray pegs you just placed in the angle beams (see Figure 2.66).

FIGURE 2.66

STEP 65 Add two regular pegs to the beam on the left and a no-friction gray peg to the beam on the right as shown in Figure 2.67.

FIGURE 2.67

STEP 66 Connect the two arms with a T-shaped beam, as seen in Figure 2.68.

FIGURE 2.68

STEP 67 Add this 3x2 cross block to the T-beam, using a cross connector (see Figure 2.69). You're done with the Mindstorms build!

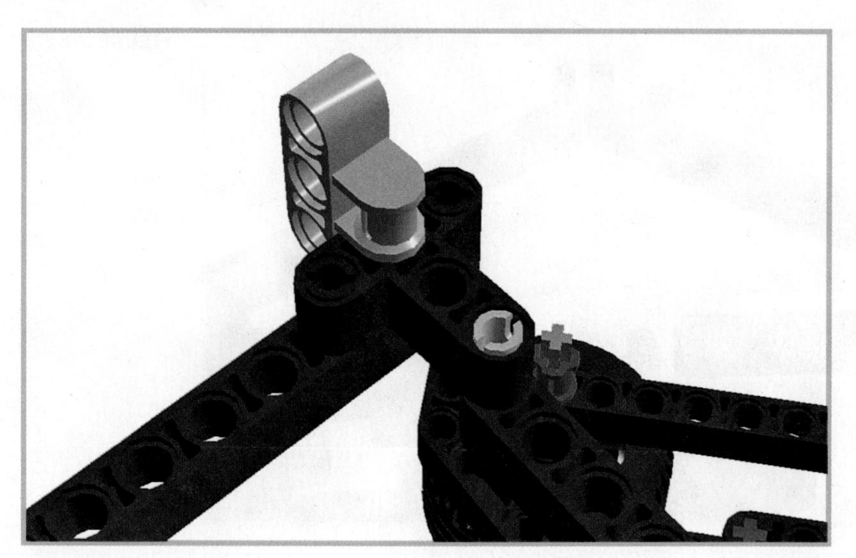

FIGURE 2.69

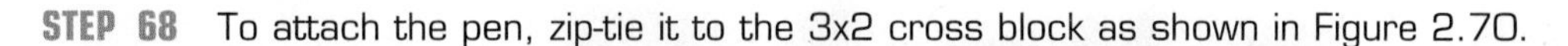

STEP 68 To attach the pen, zip-tie it to the 3x2 cross block as shown in Figure 2.70.

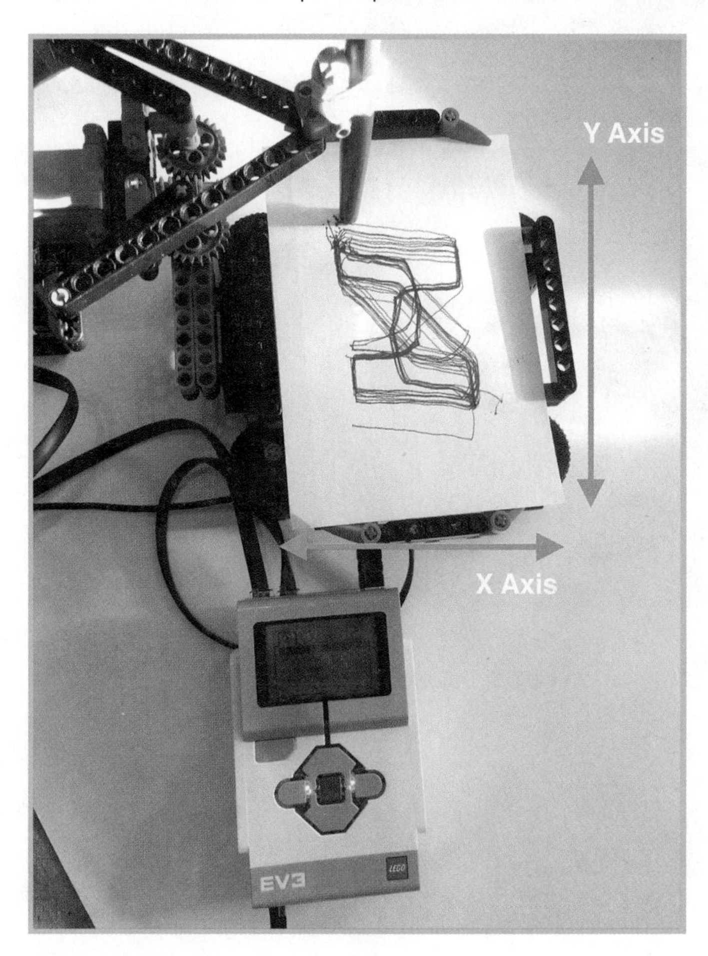

FIGURE 2.70

Program the Plotter Bot

Let's program the Plotter Bot using the EV3 Intelligent Brick. I'm assuming you are familiar with Mindstorms programming, having built the models that come with the EV3 set, so I'll skip over the basics. (If you're not, check out *Build and Program Your Own Lego Mindstorms EV3 Robots*, ISBN: 9780789751850, also from Que.)

The general principles of the program are simple: You control motor A (the small servo controlling the X axis) and motor B (the large servo moving the Y axis) to move the pen around and make shapes and lines and so on.

What exactly are X and Y axes? These are *Cartesian coordinates*, and they're a way of determining a point's location in a two-dimensional field, as depicted in Figure 2.71. There's also a Z axis (up and down), which doesn't come into play in a pen plotter, but if this were a 3D printer it certainly would!

FIGURE 2.71 X and Y help you determine a point in a two-dimensional plane.

Moving the Motors

To move the pen across the index card, you have to turn the motors' shafts. The number of rotations translates to the distance the pen moves. Here's the range I used in my plotter:

- **X axis**—The printable space is around 5.5 rotations of the servo's shaft. Because of the way the gearing is set up, the motor must turn in reverse to move the pen forward, so the range is 0 to -5.5.
- **Y axis**—The little car rolls back and forth to move the pen on the Y axis. There are only around 1.2 rotations of printable area so the range is 0 to 1.2.

Diagonal movement is achieved by turning both motors simultaneously. This presents a problem because the motors move at different speeds and have different gearing. Diagonals end up looking like curves because of this (see Figure 2.72). You can probably tweak the motors' power settings to get something of a parity in speeds, but ultimately the motors just aren't precise enough.

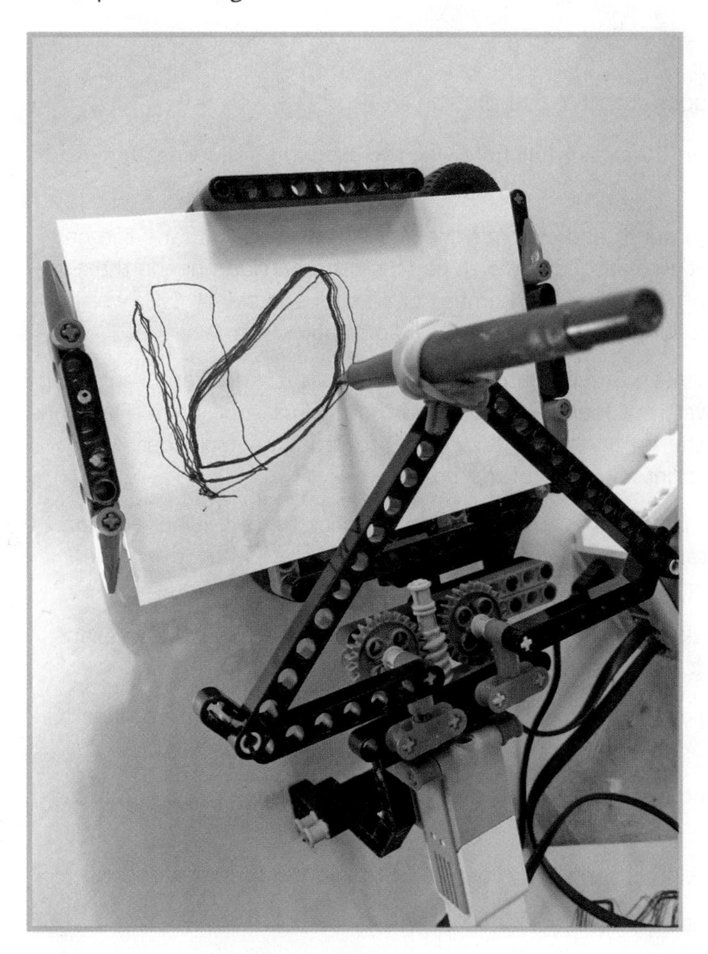

FIGURE 2.72 The speed difference between the two motors makes diagonal lines into curves.

A Simple Program

This program decorates an index card with a cross-hatch pattern of lines. It's very simple—it moves incrementally on the X axis, then the Y axis, and then repeats until the card has the lovely pattern on it.

Let's check out the basic building blocks of the program: the two motors. Figure 2.73 shows possible settings for the motors.

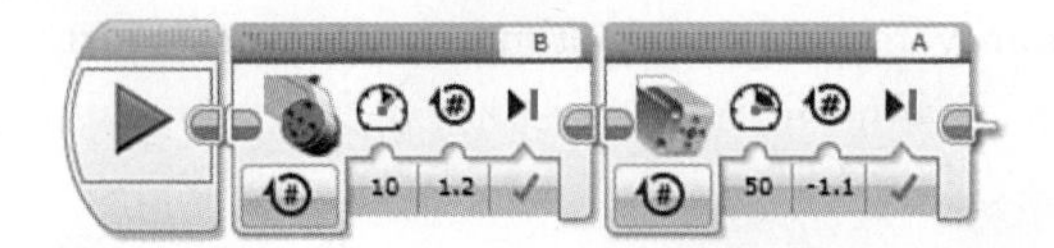

FIGURE 2.73 These two blocks control the plotter.

Here are the actions the plotter performs while following the program, following along with Figure 2.74.

STEP 1 Motor B is the large servo and controls the Y axis, essentially driving the carriage forward or backward. It's set at 10% power because the carriage needs the extra torque to keep its wheels from slipping. The number of rotations is set at 1.2, which means that it goes the entire length of the index card's printable area.

STEP 2 Motor A is the small servo and controls the X axis by extending and retracting the pen. I set it at 50% power, but feel free to change that to suit your whim. Rotations are set at -1.1. Recall that the range for the X axis is 0 to -5.5, meaning that -1.1 is one-fifth the way across the printable area of the card.

STEP 3 Add another Motor B block to move the carriage back to its original position. Note that the number of rotations is -1.2, meaning the shaft rotates backward one complete rotation plus part of another.

STEP 4 Add a Motor A block to move the X over another fifth of the width—an exact duplicate of the second block.

STEP 5 Duplicate the very first block, which moves the Y down to the bottom edge of the card's printable area.

STEP 6 Duplicate the second block to move the pen over one-fifth. The pen is now on the third fifth of the card.

STEP 7 Duplicate the third block to move the pen back to the top of Y.

STEP 8 Move the X over another fifth.

STEP 9 Move the Y down to the bottom again.

STEP 10 Move the X over another fifth. It's now at the bottom corner of the card's printable area.

STEP 11 Return the pen to X's home position.

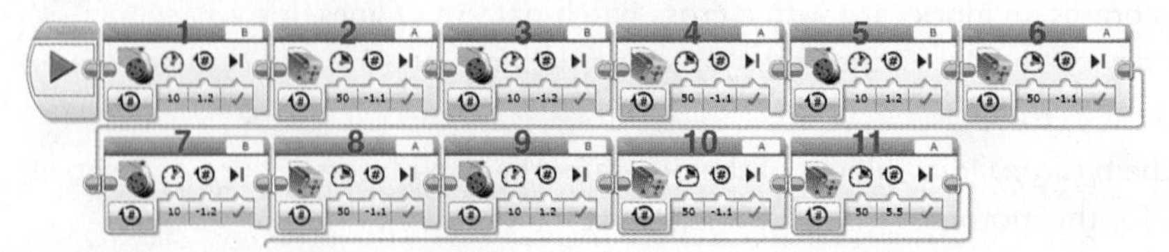

FIGURE 2.74 The first half of this program draws a series of vertical lines.

Next, let's draw the horizontal lines, basically replicating the same process as before but moving the Y incrementally while drawing horizontal lines along the X axis. Figure 2.75 shows the second half of the program.

STEP 12 The carriage returns home, moving -1.2 rotations.

STEP 13 The X axis moves -5.5 rotations.

STEP 14 Roll the carriage forward .2 rotations, or a sixth of the total Y axis.

STEP 15 Pull the X back to its home position, 5.5 rotations.

STEP 16 The Y is rolled another .2 rotations.

STEP 17 The X moves back across with -5.5 rotations, drawing another horizontal line.

STEP 18 The Y moves another 0.2 rotations.

STEP 19 The pen draws another horizontal line as the X moves 5.5 rotations.

STEP 20 The Y moves another 0.2 rotations.

STEP 21 The X moves over to the right with -5.5 rotations.

STEP 22 The Y moves 0.2 rotations.

STEP 23 The X draws its final line with 5.5 rotations. The X is now in its home position.

STEP 24 Y returns to its home position by rolling back -1.2 rotations.

This is definitely a clumsy way to program a series of movements, but it gives you the basics on controlling the Plotter Bot.

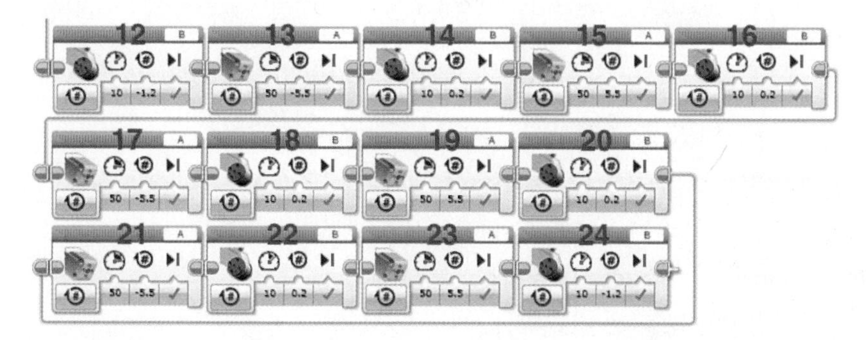

FIGURE 2.75

Resetting the X Axis

The Y axis is easy to reset—by that I mean to manually return it to its starting position. You just grab the car and put it where you want it. The X axis, however, is tricky because it has a gear assembly that doesn't like to be manually manipulated. The following is a simple program that retracts the pen to its home position, which is as close to the motor as possible.

Simply plug in the touch sensor to port 1 on the EV3 Intelligent Brick, download and run the program, and then press the button to retract the pen. Here's how to write the program, following along with Figure 2.76.

1. Drag a loop block out onto the programming canvas. The default setting (looping infinitely) is fine.
2. Drag out a switch block and put it inside the loop. Set it to Touch Sensor on port 1, with the Compare State checked. This makes the switch listen for the touch sensor, triggering the blocks in the top cell (currently empty) if the button is pressed or the bottom cell if it isn't.
3. Drag a medium motor block into the top cell and set it to 50% power and 0.25 rotations. Nothing goes into the bottom cell.

When this program is run, the X axis will retract when the button is pressed, or do nothing if it isn't. Keep this program on your EV3 brick so you can run it whenever you need to reset X.

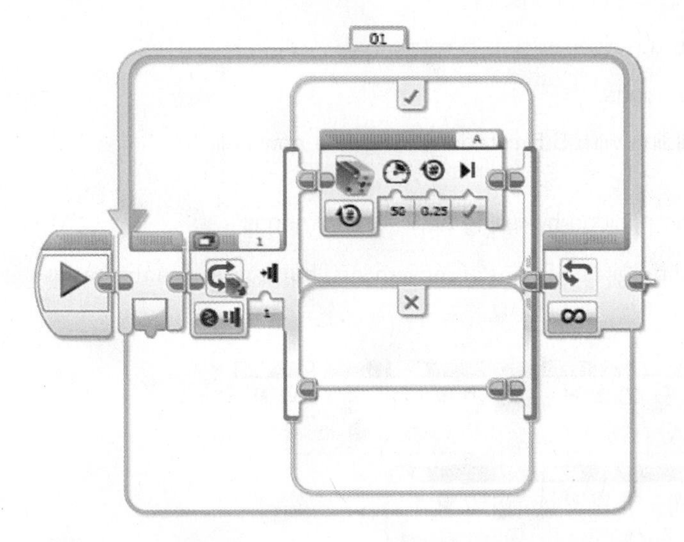

FIGURE 2.76

The X axis can be tricky to retract. This program will help.

Hacking Opportunities

In every project of this book I describe some ideas and suggestions to modify your robot. In some of them, I even include a mini project showing how to do a specific hack.

Here are some suggestions on how to make the Plotter Bot better:

- The programming is overly simplistic. You can do a lot of fun stuff with advanced programming topics such as variables and loops to greatly simplify the programming structure.
- Ideally the motors should be more precise, such as the stepper motors I described. Failing that, it may be that certain settings yield cleaner lines.
- Regular Mindstorms wheels for the Y axis would make for smoother motion than the current rolling cart's wedge-belt wheels. Even better would be some sort of rail system, but that would be hard to do with the EV3 set alone.
- Design a system that raises and lowers the pen so it doesn't have to be touching all the time.

Summary

You built your first robot, the deceptively simple Plotter Bot, in this chapter. In Chapter 3, "Hacking LEGO I: Connections," you dive into all the clever ways LEGO hackers employ to connect Intelligent Bricks to motors and sensors, and to other microcontrollers.

3

Hacking LEGO I: Connections

Mindstorms' components consist of modules linked together. The motors connect to the Intelligent Brick but not directly—they use wires to link the power and data of both modules together. Those linkages are themselves fascinating. This chapter explores what's up with those Mindstorms wires and demonstrates how to hack them into different configurations. Then, I describe some of the common methods Mindstorms hackers employ to control and connect components without using wires.

Mindstorms Wires Explained

Let's begin by exploring all the nitty-gritty details of the standard Mindstorms wire. LEGO uses semiproprietary wires in its Mindstorms variants. I call them semiproprietary because they're just a standard configuration (known in the business as RJ12) but with the tab off to one side, as shown in Figure 3.1. You literally could use RJ12s if those tabs were off-center. Since LEGO has seen fit to do it this way, however, we have to use our creativity to overcome this inconvenience.

First, however, let's check out what you get in the EV3 set:

- Four—250mm/10-inch
- Two—350mm/13.75-inch
- One—500mm/20-inch

So to recap, the EV3 set includes four short cords, one long cord, and two in the middle. The cords can be swapped end-to-end and can be used with everything from motors to sensors. They're truly universal in the Mindstorms world, meaning you only have to worry about length when you grab a wire.

FIGURE 3.1 Mindstorms cables' off-center tabs are all that differentiate them from RJ12s.

Not surprisingly, these three sizes aren't good for everyone, so some established suppliers have come up with different wire sets:

- HiTechnic's NXT Extended Connector Cable Set (P/M NWS1000) includes six cables, ranging in length from 120mm (4.7-inch) to 900mm (35.4-inch). You can buy the set at hitechnic.com.
- Mindsensors' Flexi-Cable pack (P/N FLEX-Nx) includes four cables: 200mm, 350mm, and 500mm just like regular LEGO cables. However, Mindsensors' cables have thinner and more flexible insulation, allowing them to move around and bend more readily than LEGO's stiffer wires. You can buy the Flexi-Cables at mindsensors.com.

Inside the Mindstorms Wire

So, what's going on inside that black plastic insulation? It turns out there are six smaller wires inside, as shown in Figure 3.2.

1. The blue wire is the SDA (serial data) wire, one-half of a two-wire data transfer protocol called I^2C. EV3 can transmit sensor data and commands through the I^2C bus.
2. Yellow is the SCL (serial clock) wire, the other half of the protocol.
3. Green is power, typically delivering either 3.3 or 5V from the EV3's battery pack. You can use this wire to power electronic circuitry and add-on modules.
4. Red is ground. Creating a circuit with the power pin and this ground yields 5V.
5. Black is also ground. A circuit with this ground and the power pin yields 3.3V.
6. White is analog, transmitting analog sensor signals back to the EV3 Intelligent Brick.

Knowing the purpose of each wire helps you hack them, and it never hurts to understand what's going on under the insulation.

FIGURE 3.2 The Mindstorms wire actually consists of six smaller wires.

Hacking Mindstorms Wires

Not unexpectedly, LEGO hackers have explored the wires and created their own variants to suit the needs of their projects. The following are a sampling of techniques you could employ.

Changing the Length of a Mindstorms Wire

This is an obvious one. How do you change the length of a Mindstorms wire? The following takes you through the steps, with Figure 3.3 guiding you along the process.

1. Make a shorter wire: Cut the plug off one end, making sure to leave yourself a couple of inches of wire, and trim the remaining length down to the size you want. To make a longer wire, cut an end off two wires, so that their combined length equals the size you want.
2. Carefully remove the outer black insulation and pull apart the six inner wires.
3. Solder together each wire to its same-colored mate on the other side. (If you need to polish up your soldering skills, there's a helpful how-to here: http://mightyohm.com/files/soldercomic/FullSolderComic_EN.pdf.)
4. Insulate the individual wires with heat-shrink tubing, such as SparkFun P/N 9353. Then the combined wires should get a larger piece of tubing to keep them in check.

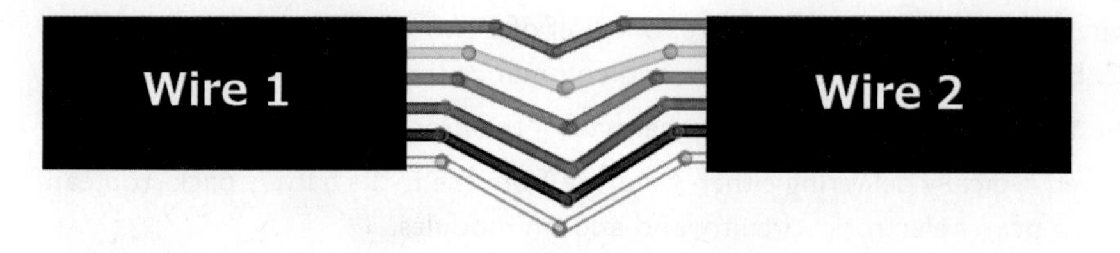

FIGURE 3.3 To alter a Mindstorms wire, just cut it apart and solder it back together.

Using a Breakout Board

Another way to access the inner workings of a Mindstorms wire is to use a breakout board. These are little circuit boards with Mindstorms-compatible plugs on them, allowing you to break out the six inner wires as separate pins.

A couple of variants are floating around; I like the Bricktronics Breakout Board, selling for only $4 from wayneandlayne.com. In Figure 3.4 I demonstrate how to light up an LED, connecting from the power pin to the red ground (5V) with a 470-ohm resistor protecting the LED from too much voltage.

FIGURE 3.4 To access the inner wires individually, use a breakout board.

Breadboard-to-PF Hybrid Wire

PF refers to Power Functions, a mostly compatible motorized set put out by LEGO and marketed alongside Mindstorms. In fact, the beams and other building elements in the EV3 set are identical to the parts sold with PF sets, making the two remarkably compatible. Not completely, however, because there is no way to control PF's awesome DC motors using your EV3 brick. Two of PF's four wires are 9V and GND, and the other two control the speed of the motor.

You still need a way to trigger the voltage—the 9V the Power Functions motors are expecting is more than the EV3 brick can handle. In Chapter 5, "Hacking LEGO II: Alternate Microcontrollers," I show you how to use an Arduino microcontroller that not only can control those great PF motors, but also Mindstorms servos as well.

In the meantime, here's how to make your own hybrid wire:

1. Cut off one end of a Power Functions extension cable (LEGO P/N 8886). It has a male end and a female end, with the male end looking like a regular 2x2 LEGO brick, and the female end looking like the underside of a similar brick, allowing you to attach them together just like they were regular bricks.
2. Strip the four individual wires on the female end. They consist of Power, Control 1, Control 2, and Ground. Solder each wire to a male header pin (SparkFun P/N 12693) or a Molex plug like you see in Figure 3.5, which you can crimp on yourself, or buy a pigtail such as the SparkFun P/N 9920. Use heat shrink to cover all conductive surfaces.

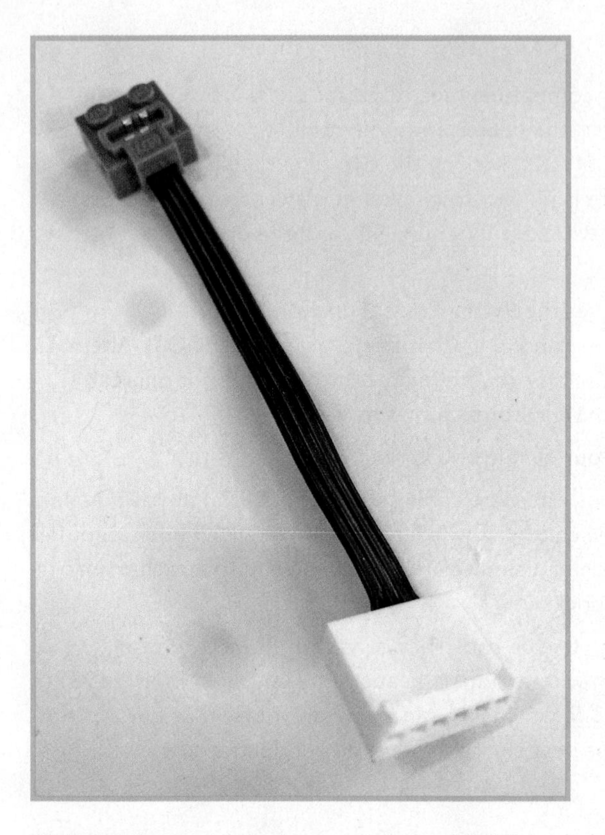

FIGURE 3.5 A cable that connects LEGO's proprietary PF connector to a breadboard.

Exploring Wireless Options

Mindstorms wires are a given—but what about wireless communication? The EV3 kit comes with two ways to communicate wirelessly with your Intelligent Brick.

Infrared Sensor and Beacon

The EV3 set includes a dirt-simple infrared remote (IR) control and receiver that allow you to control two motors on your model, both forward and backward (see Figure 3.6). In addition, you can opt between two channels, so theoretically you could control four motors with two remotes and two receivers. Another option would be to have four motors connected to your robot—for instance, two for propulsion and two to control a robot arm—and you simply switch channels when you want to do one task or the other.

The sensor has one added feature that most IR receivers lack: It can be used as a proximity sensor, beaming out infrared light and sensing as it bounces back. This feature has a short range compared to other proximity sensors (for example, ultrasonic), and can detect proximity only within 50cm to 70cm, or around 2 to 3 inches.

The beacon is what LEGO calls its remote control, and this is not just for fun: One of the projects described in the EV3 set is an IR-homing robot that wanders around until it senses the infrared signal from the beacon and rolls toward it. The controller's range is only about 2 meters, unfortunately.

FIGURE 3.6 The infrared sensor and beacon give you simple wireless control of your robot.

Bluetooth

Another intriguing option is the EV3 brick's Bluetooth capabilities. The Intelligent Brick has a Bluetooth chip on-board, allowing it to connect to other EV3 bricks as well as take commands from smartphones using an application called the Commander, which includes preset control configurations for the five sample robots that are part of the EV3 set (see Figure 3.7). You can also create an interface for a custom robot, pulling out sliders and buttons from a library to match what you're building.

EV3's Bluetooth capability also allows you to control the Intelligent Brick from your PC or Mac wirelessly, just as if you had it plugged in with a Bluetooth cable.

Finally, one cool aspect of the robot, both in terms of Bluetooth and regular wiring, is you can link up to four EV3 bricks together if you want to build a gloriously complicated robot.

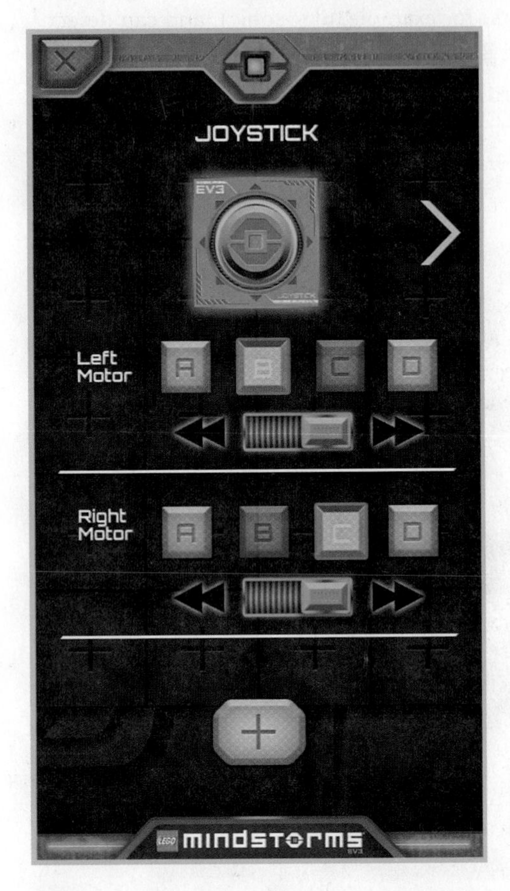

FIGURE 3.7 LEGO's Bluetooth app allows you to control robots wirelessly.

Hacking Wireless

It almost goes without saying that Mindstorms fans have figured out how to control their robots in ways not officially supported by LEGO. Here are just a few ways to wirelessly control your Mindstorms robot.

XBee

A common hobbyist and professional wireless specification is called Zigbee, and XBee is a brand of wireless modules built to that spec. Dexter Industries (dexterindustries.com) sells a

Mindstorms-compatible XBee breakout called the NXTBee, though I'm not sure whether it's compatible with EV3 yet. Another technique is to ditch the EV3 brick altogether and use an Arduino: Check out the cool LEGO bracer shown in Figure 3.8. It has an Arduino, battery pack, XBee, and Wii nunchuk, allowing me to operate a robot with a wearable controller. SparkFun sells XBee radios (P/N 8665) as well as its own flavor of breakout board.

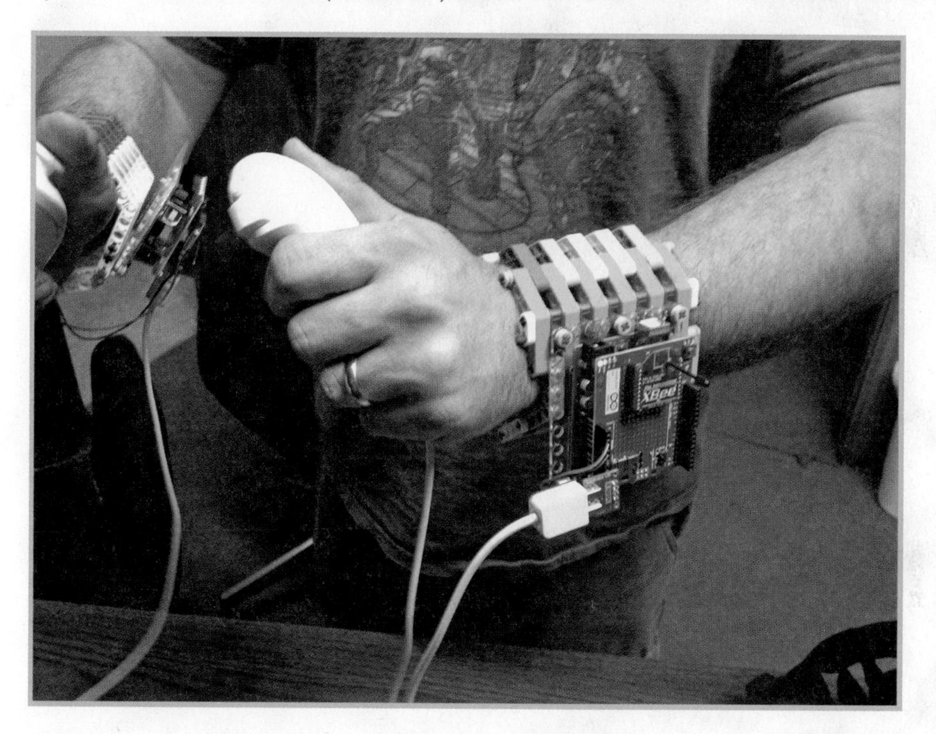

FIGURE 3.8 This wireless controller combines LEGO, Wii, and Arduino.

Radio Control

Normal radio control (RC) technology doesn't mesh well with Mindstorms, but it can be made to work. RC flight electronics consist (in their most basic configuration) of a radio, shown in Figure 3.9, as well as a receiver. The receiver interprets the data from the transmitter and triggers pins that tell the motors what to do. Not surprisingly, those same pins can trigger Arduino actions or could be used to bump Mindstorms touch sensors with a servo.

FIGURE 3.9 An RC transmitter and receiver can control Mindstorms models.

PlayStation Controller

Mindsensors.com and a couple of other places sell a wireless controller that consists of a PlayStation 2 (PS2) interface card that plugs into the EV3 Intelligent Brick—onto which a wired PS2 controller may connect (see Figure 3.10). Mindsensors also sells a 2.4Ghz wireless PS2 controller and a matching dongle that plugs into that interface card, allowing you to wirelessly control your robot.

FIGURE 3.10 Mindsensors' PS2 adapter lets you control your model with a game controller.

Wi-Fi Dongle in EV3

There is no native Wi-Fi capability in EV3 bricks, but you can add it with a USB dongle, such as the NetGear WNA1100 shown in Figure 3.11. As a matter of fact, the WNA1100 is currently the *only* wireless dongle that the EV3 works with out of the box. It may be that other models can be made to work with the EV3, but so far just this one works.

FIGURE 3.11 The NetGear WNA1100 is the only Wi-Fi dongle that works with the EV3.

BrickPi and a Wi-Fi Module

Here's another example of a Wi-Fi add-on module allowing wireless communication of a Mindstorms robot. The BrickPi shield allows you to control Mindstorms by doing away with the EV3 brick and using a Raspberry Pi minicomputer, with the BrickPi mounted on top (see Figure 3.12). A Wi-Fi module from Adafruit (P/N 814) provides connectivity, though the Pi's built-in Ethernet port is always an option.

FIGURE 3.12 The BrickPi shield helps control Mindstorms robots.

Summary

This chapter is all about connections: hacking Mindstorms wires and playing around with wireless options such as infrared, radio control, and Wi-Fi. In Chapter 4, "Project: Remote-Controlled Crane," you put this knowledge to good use, making a rolling crane that responds to a variety of wireless control methods.

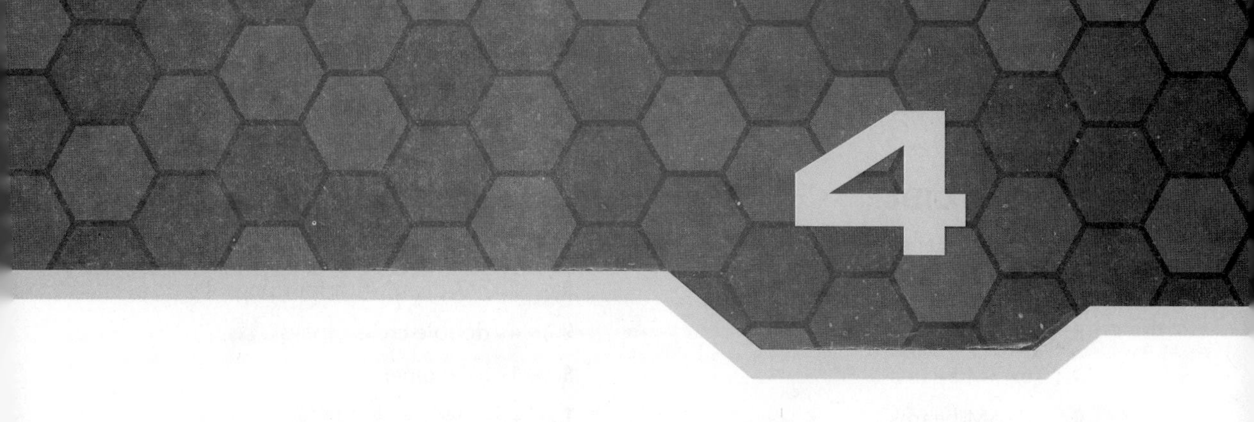

Project: Remote-Controlled Crane

The next project teaches you two different ways to wirelessly control a robot. You build a Remote-Controlled Crane that rolls along a tabletop or shelf and lowers a line with a basket or hook attached (see Figure 4.1). After the build is done, you program the crane to accept commands from the EV3 remote control, allowing you to remotely operate the robot's three motors. Finally, you learn how to set up your smartphone to control the robot, taking advantage of the EV3's Bluetooth capability.

FIGURE 4.1 The Remote-Controlled Crane may be controlled two different ways.

Parts List

Gather together the parts shown in Figure 4.2, all found in the LEGO Mindstorms EV3 set:

A. 3x 15M beams

B. 1x 13M beam

C. 6x 7M beams

D. 3x 5M beams

E. 3x 3M beams.

F. 2x double-angle beams

G 3x 3x5 90-degree angle beams

H. 4x 2x4 90-degree angle beams

I. 4x T-beams

J. 4x wheels and rims

K. 1x 36-tooth gear

L.. 2x wedge-belt wheels

M. 2x hubs

N. Bushings: You need both regular and half bushings. Just grab 'em all.

O. 2x 5x11 frame bricks

P. 2x 5x7 frame bricks

Q. 2x 3M beam with four pegs

R. 4x double cross blocks

S. 1x axle joiner

T. 4x cross block

U. 1x 0-degree angle element

V. 9M cross axle

W. 8M cross axle with end stop

X. 6M cross axle

Y. 1x 5M cross axle

Z. 3x 3M cross axles

AA. Pegs: You need all kinds. Just grab 'em all!

BB. 1x EV3 Intelligent Brick

CC. 1x IR sensor

DD. 2x large motors

EE. 1x medium motor

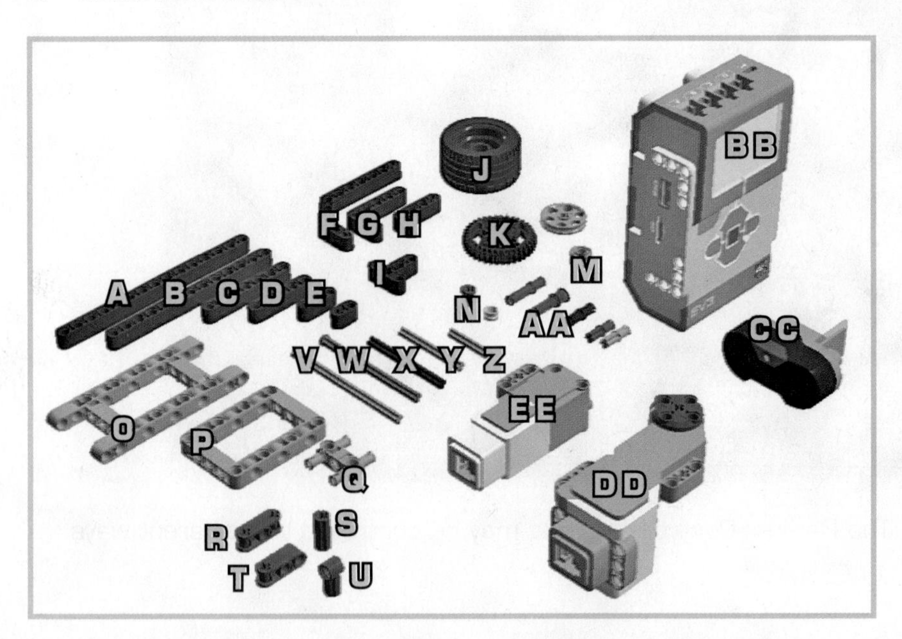

FIGURE 4.2 You need these parts to build the Remote-Controlled Crane.

Building the Crane

This is a complicated build! Grab your parts and get started.

STEP 1 Use a connector peg to attach a double cross block to a frame brick, as shown in Figure 4.3.

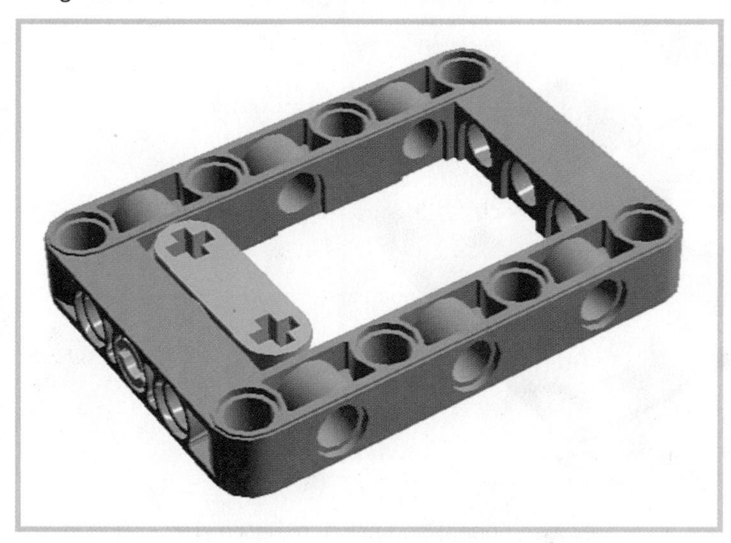

FIGURE 4.3

STEP 2 Insert four pegs, two each of the blue cross connectors and black regular ones. As one might surmise, the cross connectors go into the cross-shaped holes and the regular connectors go into the round ones.

Figure 4.4 shows how it should look.

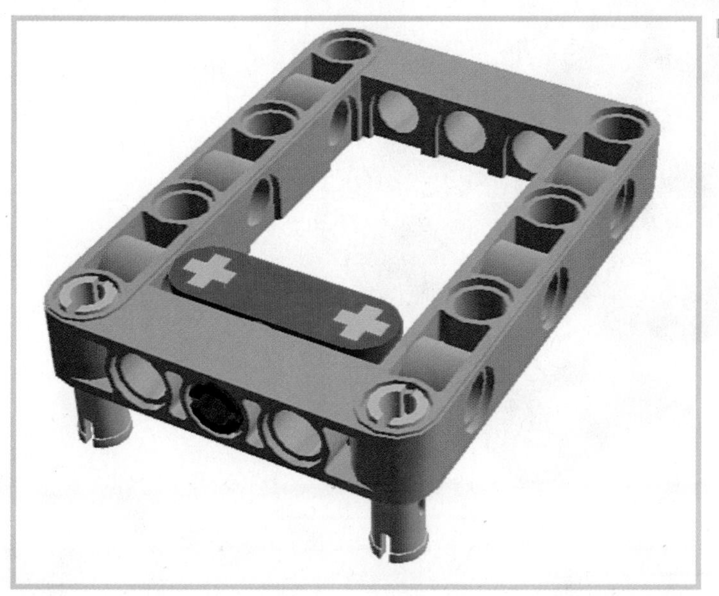

FIGURE 4.4

STEP 3 Add a pair of angle beams to the pegs, and add a pair of cross pegs in addition (see Figure 4.5).

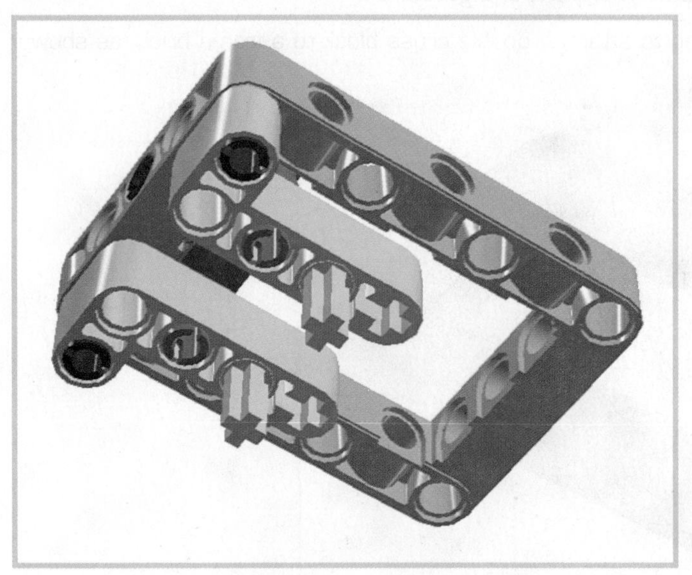

FIGURE 4.5

STEP 4 Add a 36-tooth gear to the exposed pegs, as shown in Figure 4.6.

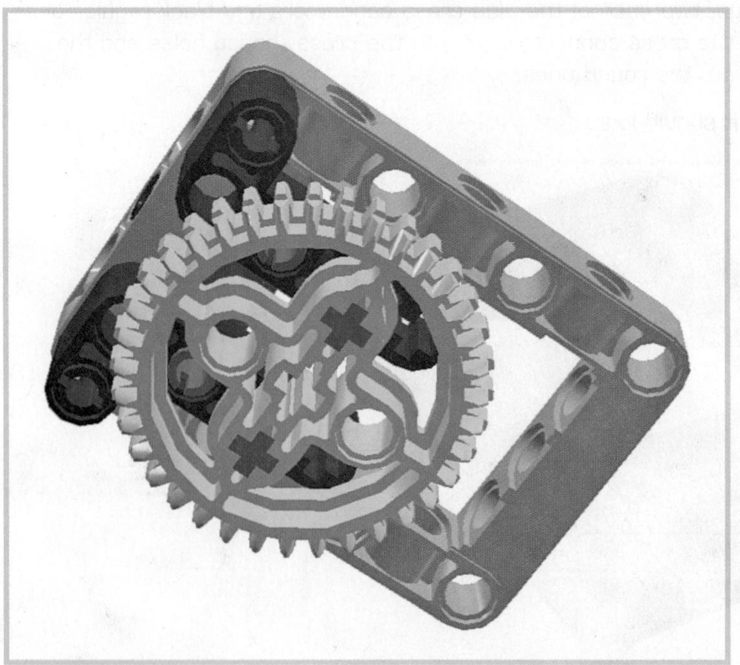

FIGURE 4.6

STEP 5 Flip over the assembly and insert four 3M connector pegs. It should look like Figure 4.7.

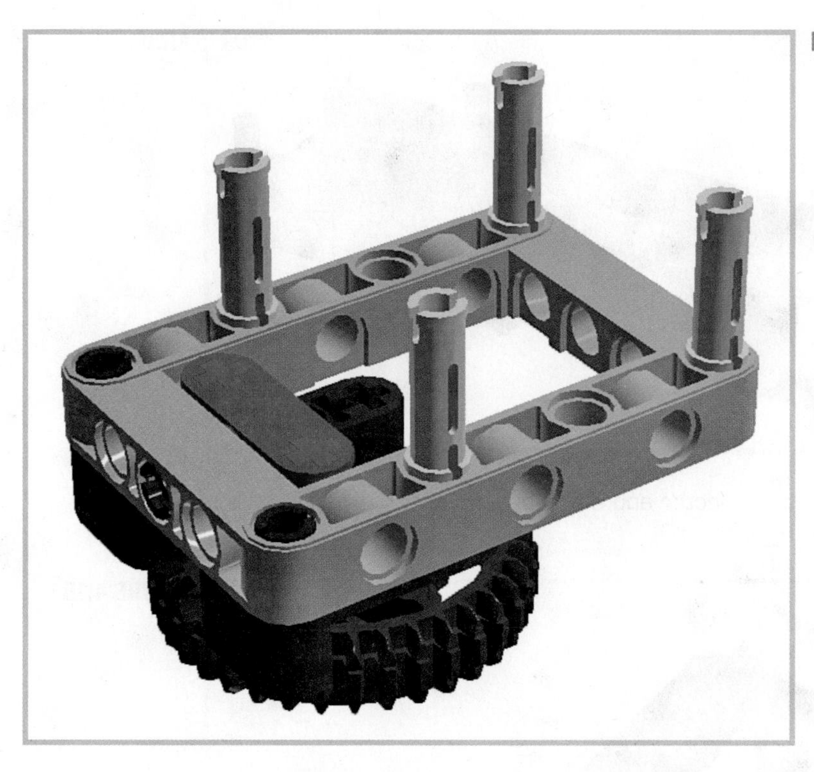

FIGURE 4.7

STEP 6 Attach a pair of 15M beams, and while you're at it, add six connector pegs to the undersides, as shown in Figure 4.8.

FIGURE 4.8

STEP 7 Grab a 5x11 frame brick and attach it to the assembly (see Figure 4.9). The crane is starting to take shape!

FIGURE 4.9

STEP 8 Add a cross connector and a 2M beam to the frame brick. This loop helps guide the cord (see Figure 4.10).

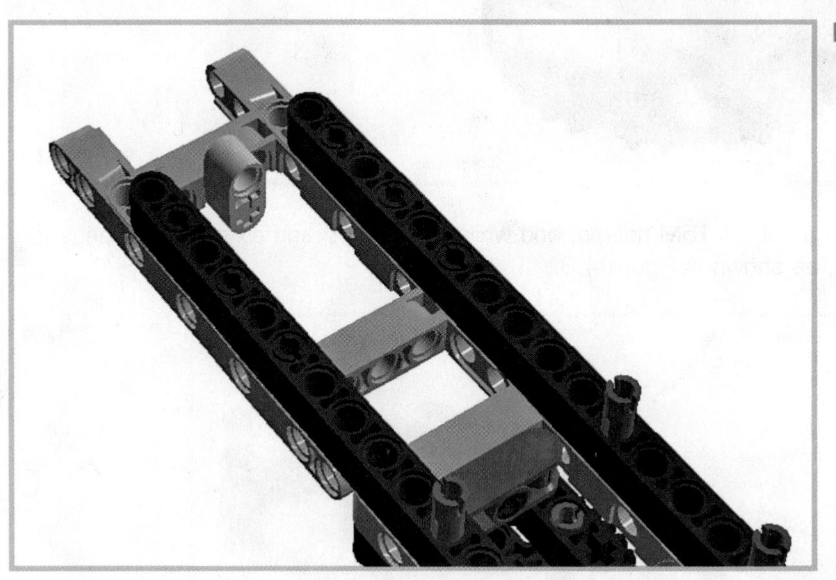

FIGURE 4.10

STEP 9 Add another 5x11 frame brick to the ends of those four 3M pegs you added way back in step 5. Figure 4.11 shows how it should look.

FIGURE 4.11

STEP 10 Attach the medium motor to the frame brick using a 5M cross axle and a pair of 2M pegs with cross hole. These pegs are great if you can't get a regular connector peg in there. It's like a peg with a bushing on one end. In Figure 4.12 I show the motor with the pegs only partially inserted. Plug them in all the way to secure the motor!

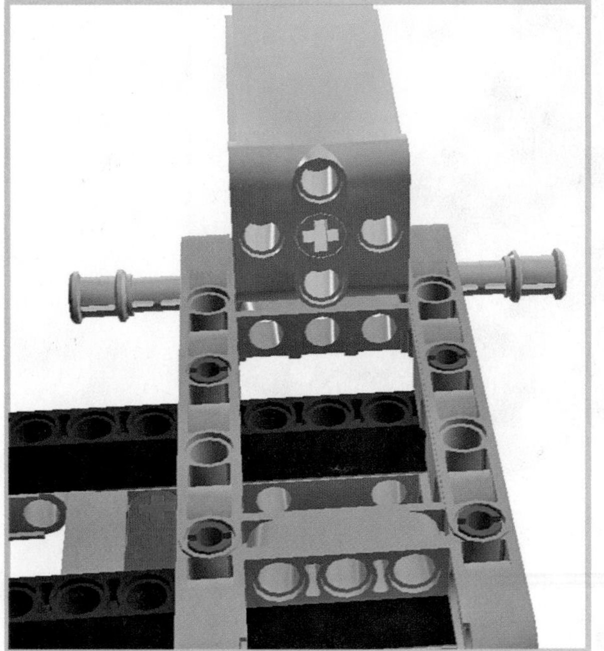

FIGURE 4.12

STEP 11 Insert a pair of connector pegs as shown in Figure 4.13.

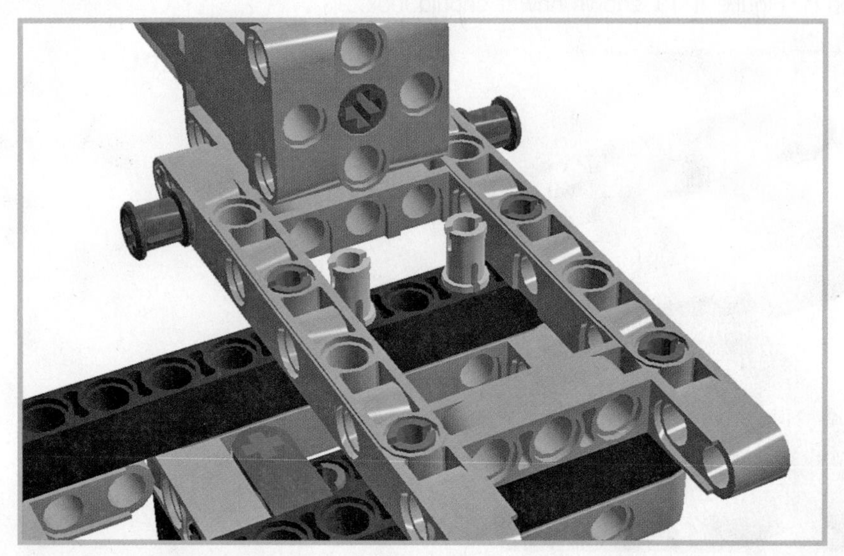

FIGURE 4.13

STEP 12 Next, add a 3M beam and insert a cross connector into the free center hole (see Figure 4.14).

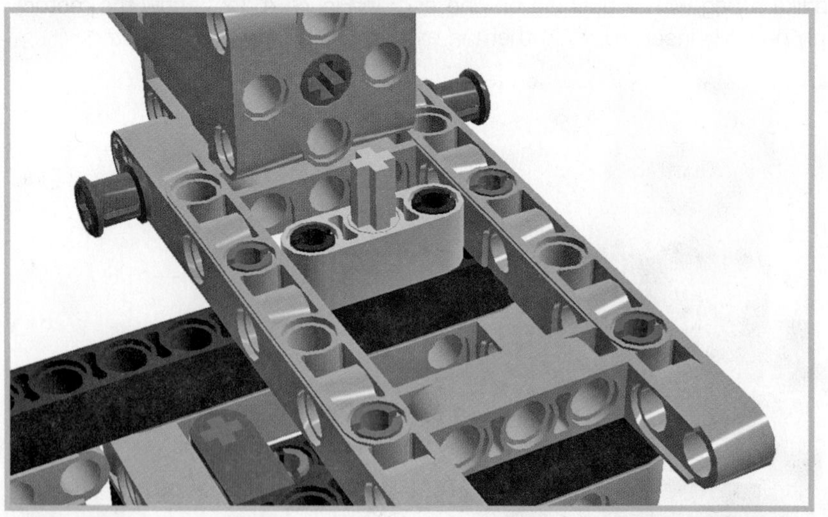

FIGURE 4.14

STEP 13 Add a 0-degree angle element to the cross connector, as shown in Figure 4.15.

FIGURE 4.15

STEP 14 Insert a 6M cross axle into the motor's hub, threading it through the angle element as well as a pair of bushings. It should look like Figure 4.16.

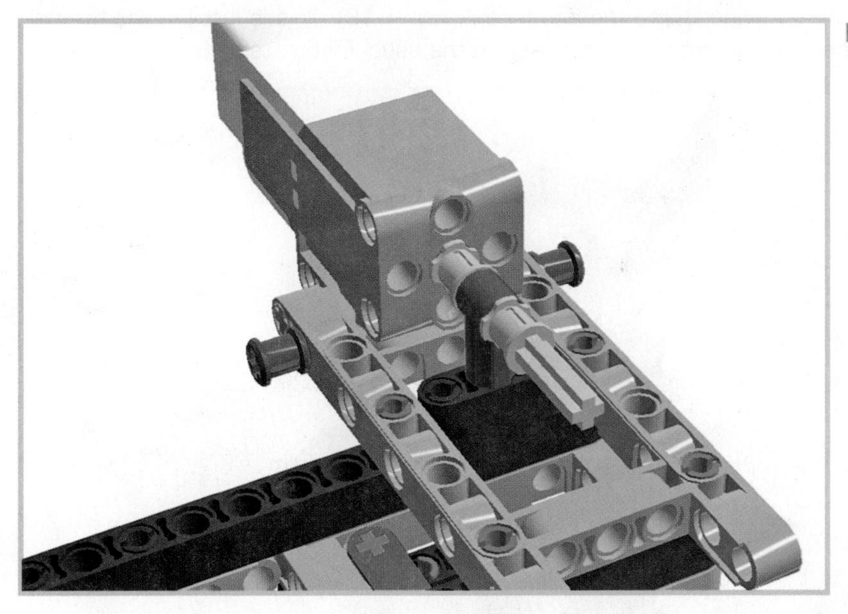

FIGURE 4.16

STEP 15 Add a pair of wedge-belt wheels to the axle (see Figure 4.17).

FIGURE 4.17

STEP 16 The end of the axle is secured with an axle joiner, with a nonfriction cross connector (they're beige) plugged into the end (see Figure 4.18). This is just like the normal blue cross connectors but rotates freely, lacking the usual friction tabs.

FIGURE 4.18

STEP 17 Add three connector pegs as shown in Figure 4.19.

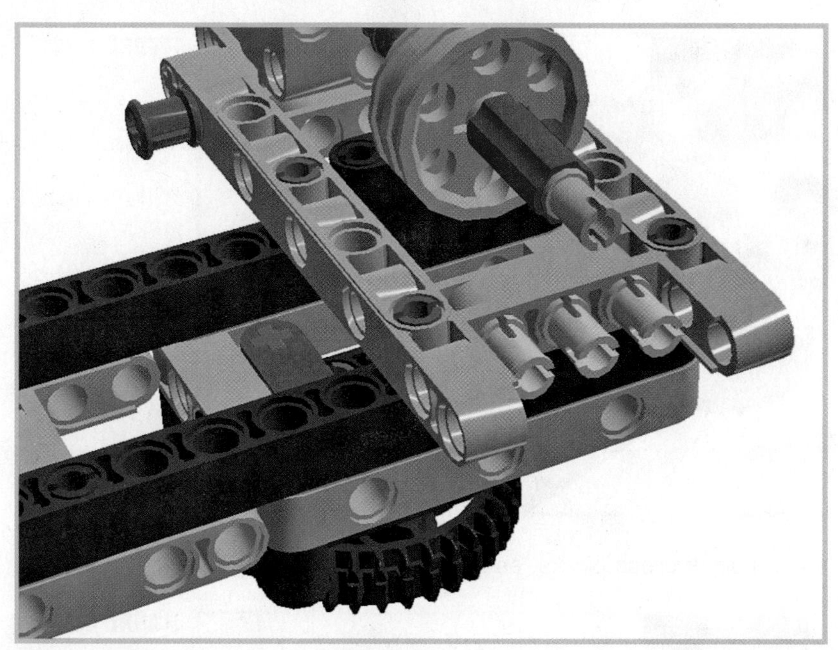

FIGURE 4.19

STEP 18 Complete the axle assembly with a T-beam that attaches to the exposed pegs (see Figure 4.20).

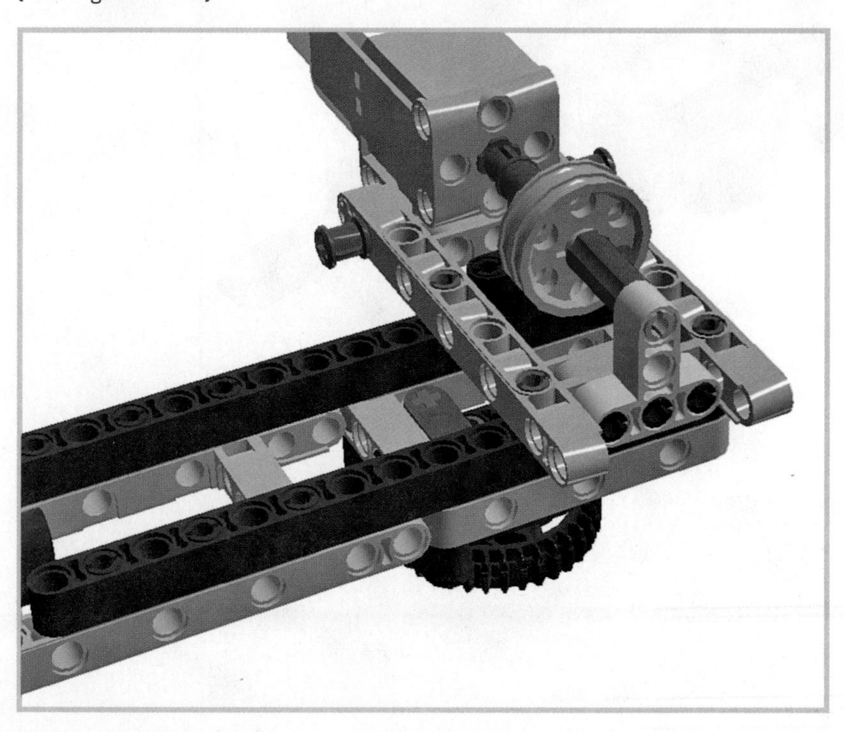

FIGURE 4.20

STEP 19 Add four cross connectors as shown in Figure 4.21.

FIGURE 4.21

STEP 20 Next, add a pair of cross blocks, as shown in Figure 4.22.

FIGURE 4.22

STEP 21 Let's complete the crane portion of the build with this pair of rolling assemblies; they consist of 6M cross axles with bushings and small hubs. Figure 4.23 shows how it should look. When you're done, set aside the crane for now, while you work on the rolling platform, which I call the "car."

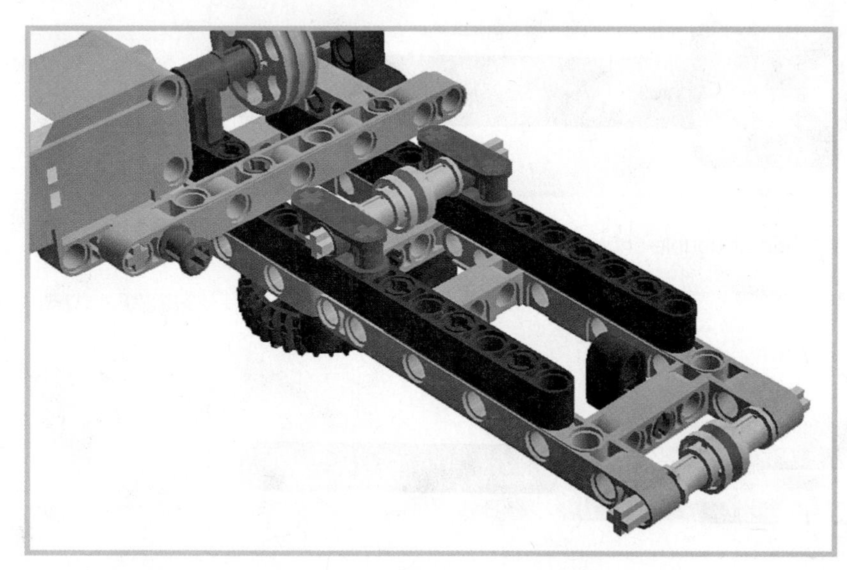

FIGURE 4.23

STEP 22 Let's start with the car part of the build. Begin with a 15M beam, and then add a pair of 3M connectors to it (see Figure 4.24).

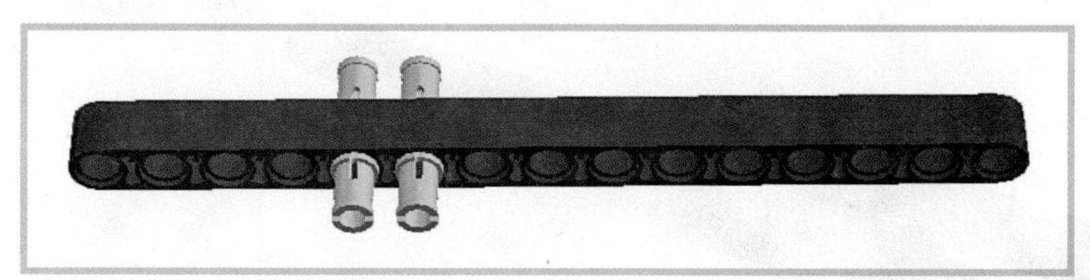

FIGURE 4.24

STEP 23 Add a pair of cross blocks to the 3M pegs, as shown in Figure 4.25.

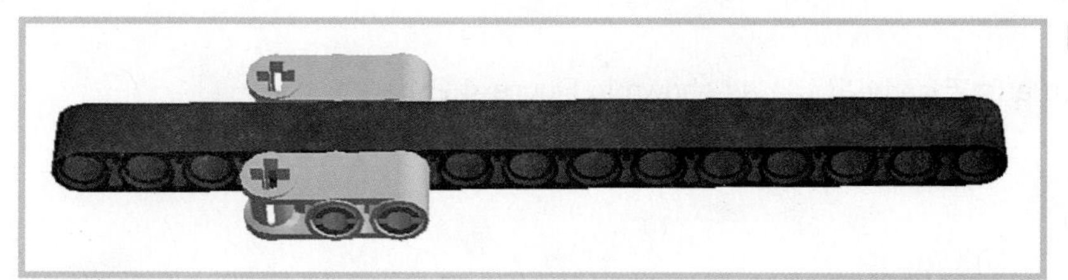

FIGURE 4.25

STEP 24 Insert two 3M pegs and add two cross connectors to the cross blocks. It should look like Figure 4.26.

FIGURE 4.26

STEP 25 Attach a pair of double-angle beams, as shown in Figure 4.27.

FIGURE 4.27

STEP 26 Add a 3M beam with pegs to the cross connectors (see Figure 4.28).

FIGURE 4.28

STEP 27 Attach a 5x7 frame brick, as shown in Figure 4.29.

FIGURE 4.29

STEP 28 Attach a 7M beam along with a pair of connector pegs (see Figure 4.30).

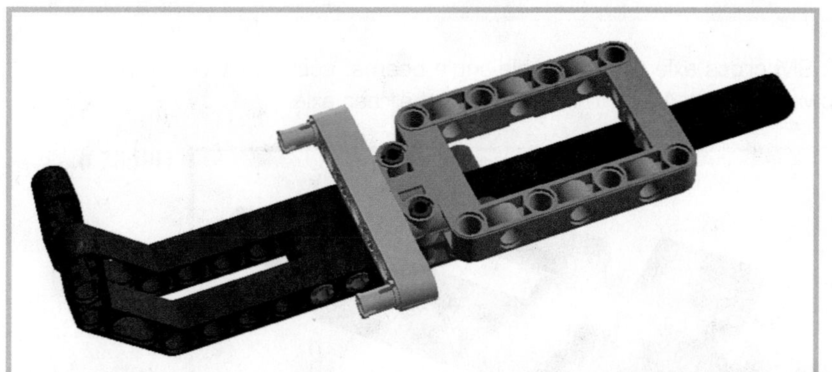

FIGURE 4.30

STEP 29 Another 7M beam is added, along with two more pegs. The assembly should look like Figure 4.31.

FIGURE 4.31

STEP 30 Add yet another 7M beam with some more pegs—though this time, the pegs are not arranged symmetrically (see Figure 4.32).

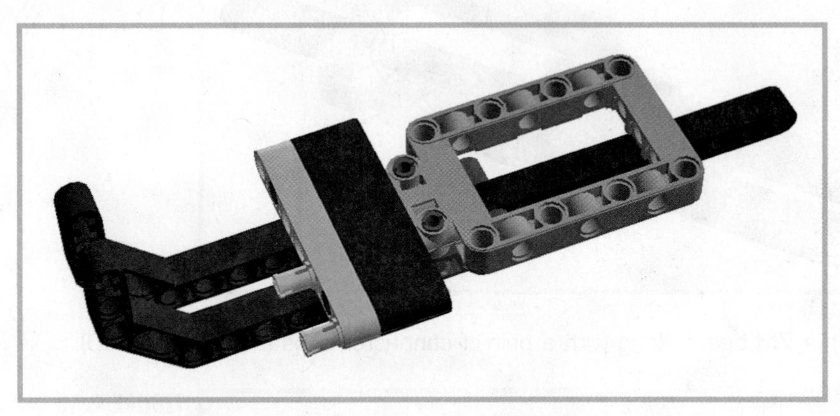

FIGURE 4.32

STEP 31 Add a 9M cross axle to the double-angle beams, securing it with three bushings as shown in Figure 4.33. This serves as the rear axle.

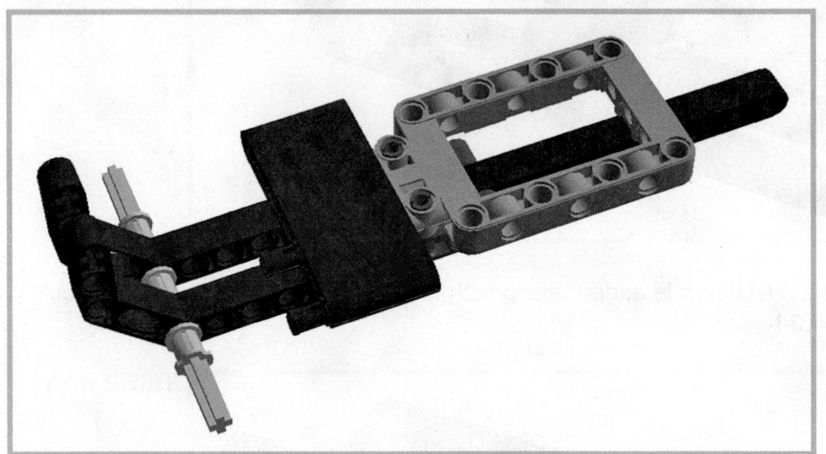

FIGURE 4.33

STEP 32 Attach the rims and tires, securing the ends with bushings as shown in Figure 4.34.

FIGURE 4.34

STEP 33 Let's switch things up a tiny bit by adding a pair of beams to the EV3 Intelligent Brick (see Figure 4.35). The beams are 5M and 7M.

FIGURE 4.35

STEP 34 Attach the beams on the Intelligent Brick to the 7M beams on the chassis (see Figure 4.36). It will be a little wobbly; don't worry! You secure the other side in the next few steps.

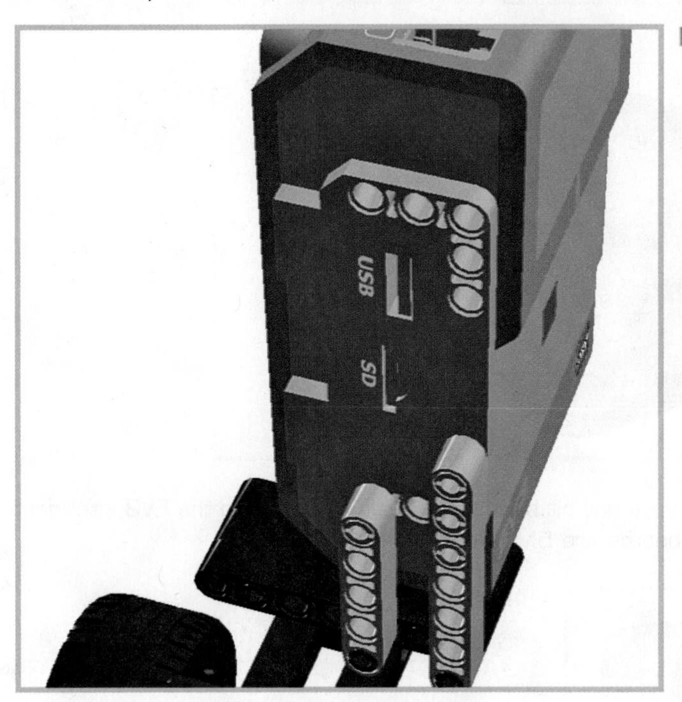

FIGURE 4.36

STEP 35 Insert four connector pegs as shown in Figure 4.37.

FIGURE 4.37

STEP 36 Next, attach a 3x5 angle beam to the pegs you just added (see Figure 4.38).

FIGURE 4.38

STEP 37 Add the motor connection hardware: a 3M beam with pegs and a trio of regular connector pegs. It should look like Figure 4.39.

FIGURE 4.39

STEP 38 Attach a large motor to the pegs you just added (see Figure 4.40). In addition, a 3M cross axle connects the motor to the 15M beam.

FIGURE 4.40

STEP 39 Slide a rim onto an 8M cross axle with end stop and add a half bushing. Slide this assembly onto the motor's hub and then secure the other end with another half bushing. Figure 4.41 shows how it should look.

FIGURE 4.41

STEP 40 Add the other wheel and secure with a bushing, as shown in Figure 4.42.

FIGURE 4.42

STEP 41 Use a 3M cross axle to attach two cross blocks to the motor (see Figure 4.43).

FIGURE 4.43

STEP 42 Add six connector pegs as shown in Figure 4.44.

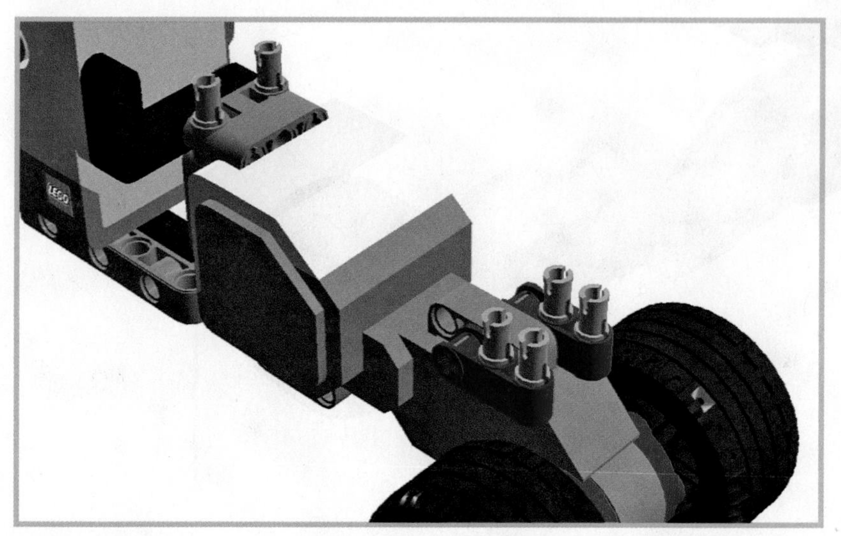

FIGURE 4.44

STEP 43 Add a pair of T-beams. It should look like Figure 4.45.

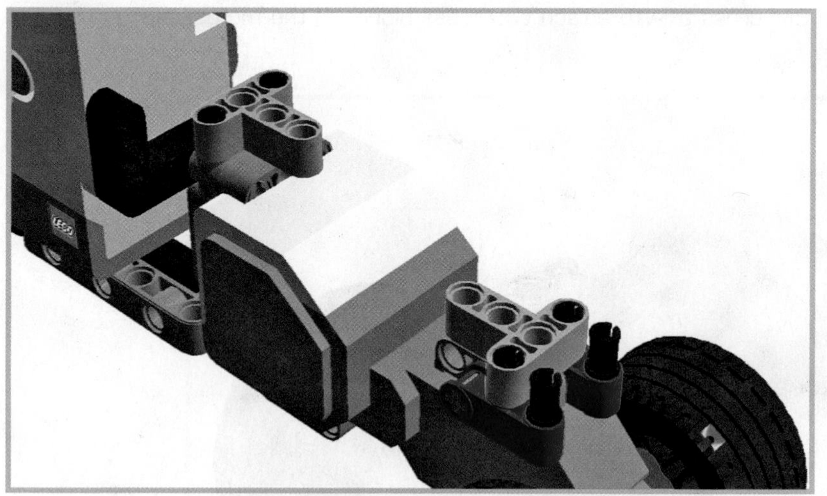

FIGURE 4.45

STEP 44 Next, insert three connector pegs and add a 3M beam, as shown in Figure 4.46.

FIGURE 4.46

STEP 45 Add a 7M beam as well as four 3M connector pegs (see Figure 4.47).

FIGURE 4.47

STEP 46 Add a pair of 3x5 angle beams, as shown in Figure 4.48. Note that a single connector peg is added as well.

FIGURE 4.48

STEP 47 This next step is tricky. Connect a 7M beam to a 2x4 angle beam with the help of a cross connector. Attach this to the free connector peg you added in the previous step. It should look like Figure 4.49.

FIGURE 4.49

STEP 48 Insert six connector pegs as you see in Figure 4.50.

FIGURE 4.50

STEP 49 Add a 13M beam to that glorious row of pegs (see Figure 4.51). That is one secure beam!

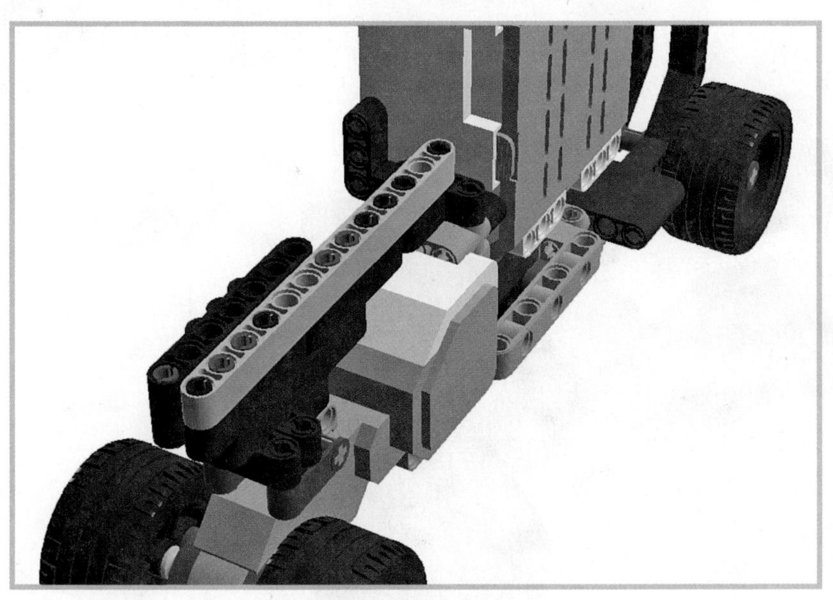

FIGURE 4.51

STEP 50 Insert two connector pegs and a cross connector to the angle beams, as shown in Figure 4.52. Then use another peg to connect a 3M beam to the 7M beam.

FIGURE 4.52

STEP 51 Insert a pair of 3M connector pegs, as shown in Figure 4.53.

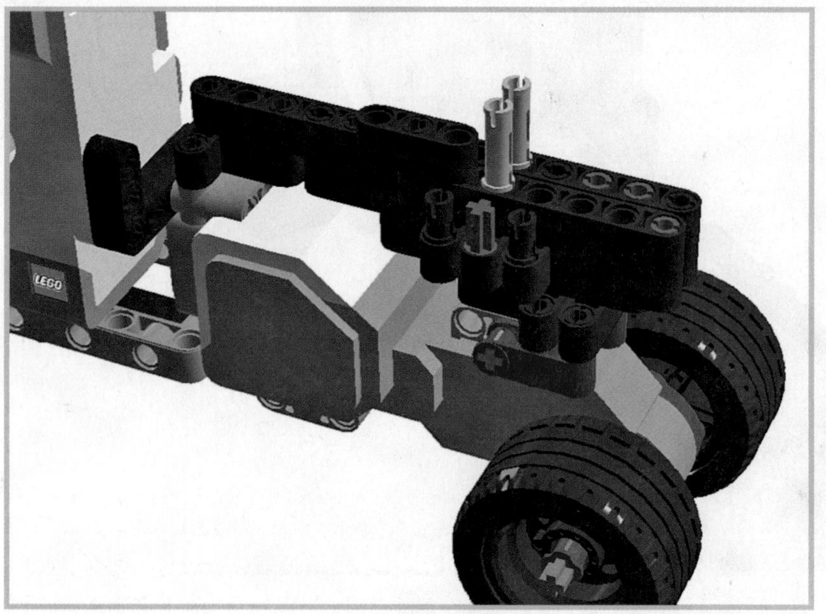

FIGURE 4.53

STEP 52 Here is another tricky step. Slide a T-beam between the motor's two mounting rails, holding it into place with your fingers. Attach the motor to those three pegs you added in step 50, and the bottom of the T-beam connects to the two 3M pegs. Figure 4.54 shows how it should look.

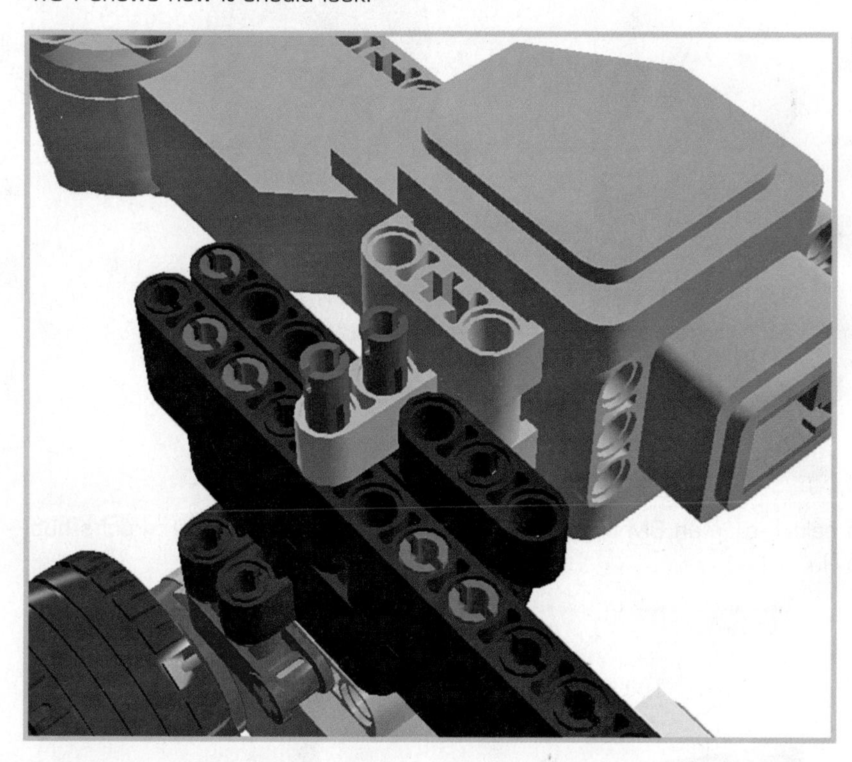

FIGURE 4.54

STEP 53 Insert five 3M connector pegs as shown in Figure 4.55, securing a 2x4 angle beam along the way.

FIGURE 4.55

STEP 54 Top off the motor mount with a pair of 5M beams, as shown in Figure 4.56.

FIGURE 4.56

STEP 55 From below, slide an 8M cross axle with end stop up through the motor's hub (see Figure 4.57).

FIGURE 4.57

STEP 56 Next up, attach the IR sensor. Begin by inserting two cross connectors, as shown in Figure 4.58.

FIGURE 4.58

STEP 57 Attach a cross block to the two pegs and add another cross connector to the block's hole. Figure 4.59 shows how it should look.

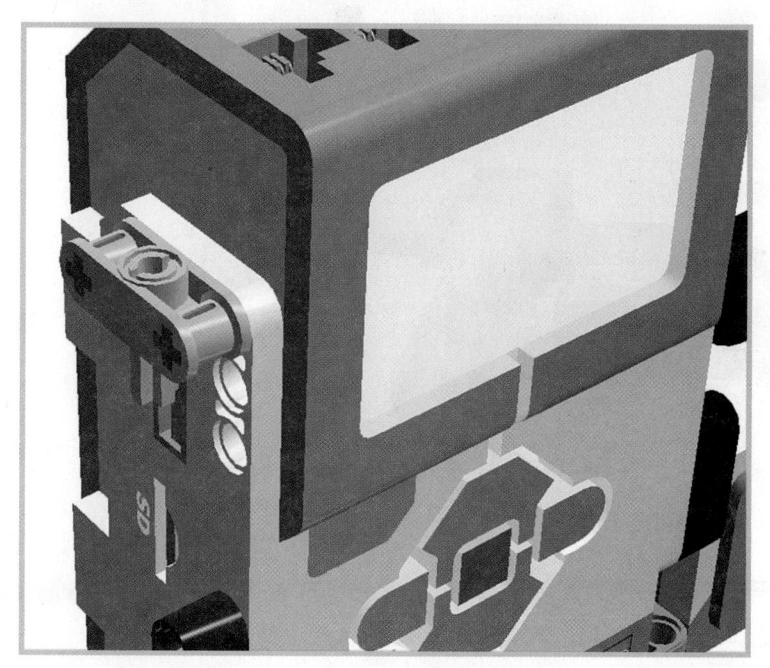

FIGURE 4.59

STEP 58 Attach the IR sensor to the cross block as shown in Figure 4.60.

FIGURE 4.60

STEP 59 Grab the crane arm you built already and mount it to the 8M axle with end stop you added in Step 55. Slide the axle through the center mounting hole of the gear. You're done (see Figure 4.61)!

FIGURE 4.61

Programming the Crane

If you've built and programmed all the robots that come with the EV3 set, chances are you'll find this Mindstorms program easy. It's really a cinch, so let's get started.

STEP 1 Drag out a Loop Block, as shown in Figure 4.62. The default settings are fine. The Loop ensures the robot continually listens (or maybe "looks" since it's IR?) for commands.

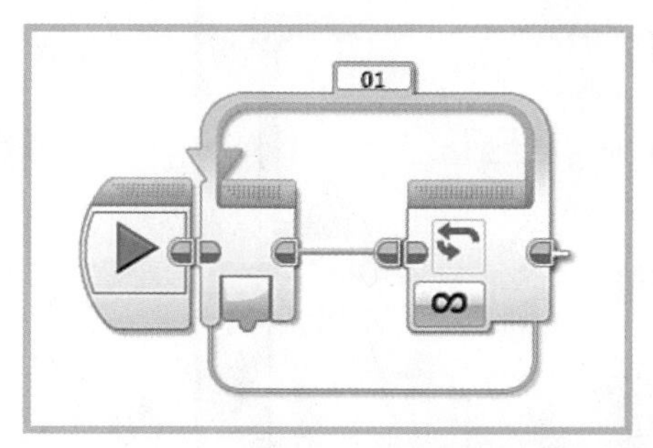

FIGURE 4.62

STEP 2 Grab a Switch Block and put it inside the Loop, as shown in Figure 4.63. Set the Switch to Infrared Sensor - Measure - Remote. Notice that it shows two button configurations for the two cells of the Switch Block: The top shows a single button pressed (the red dot) and the bottom shows no button pressed. These are merely the default options—we can and will change them. Finally, ensure that the channel is set to 1.

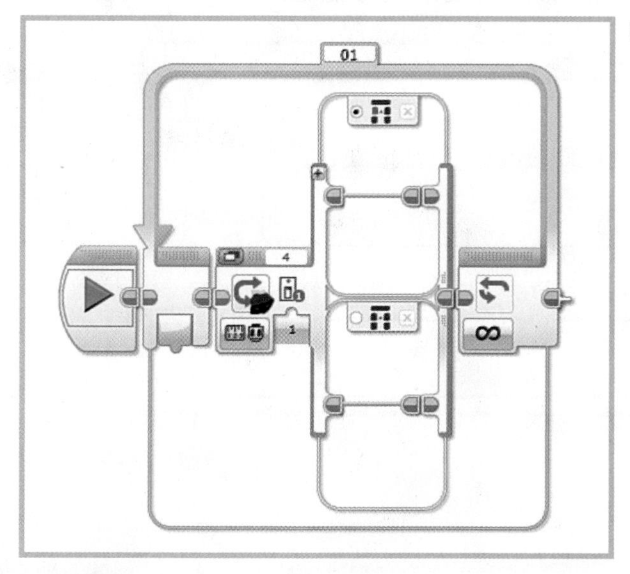

FIGURE 4.63

STEP 3 Click on the black and white icon in the top-left corner of the Switch Block to change it to tabbed view. You may need to drag out the boundaries of your window to see everything. Start making tabs by hitting the + sign on the tab bar. You need seven total tabs, one for each action. It should look like Figure 4.64. The seven tabs represent seven possible actions the robot can perform: forward and backward for three motors, and nothing (the robot just sits there).

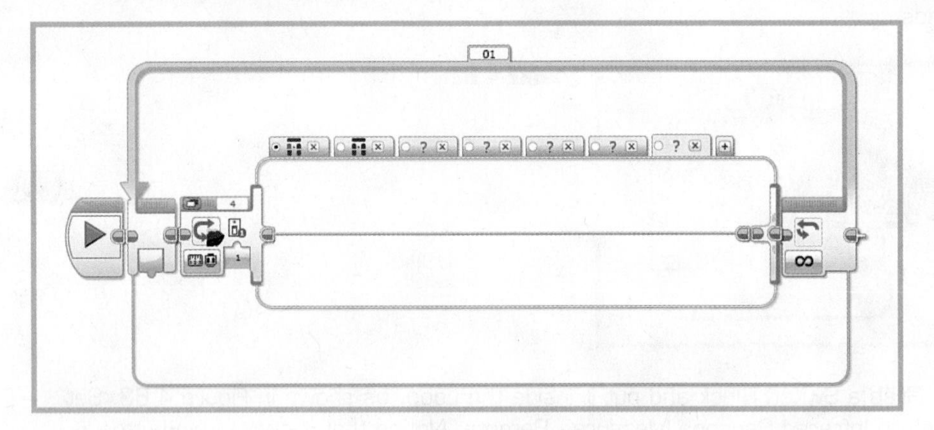

FIGURE 4.64

STEP 4 Set the first tab so that no buttons are red (that is, pressed) and make it the default by clicking on the circle. There is no action for this tab, as we want the motors to be idle until a button is pressed. You can see how it should look in Figure 4.65.

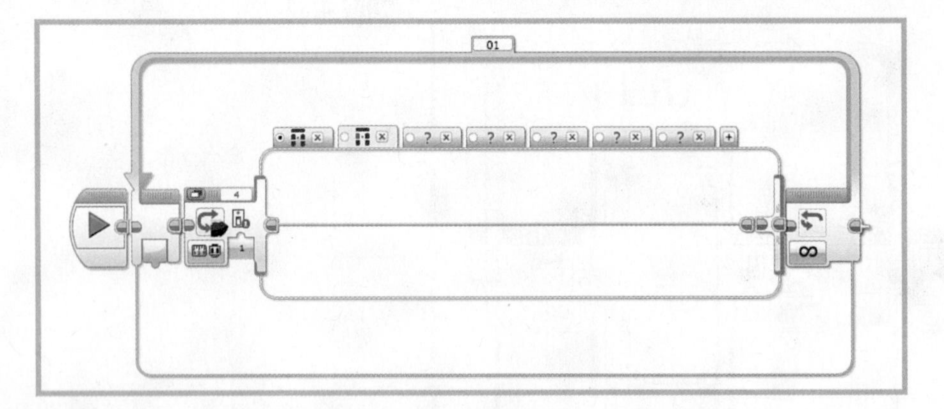

FIGURE 4.65

STEP 5 Give each remaining tab a unique button designation. You can see in Figure 4.66 that I use the left-hand pair of buttons to control one motor (forward and reverse), the right-hand pair of buttons to control a second motor, and finally, a double-button press to control the third motor.

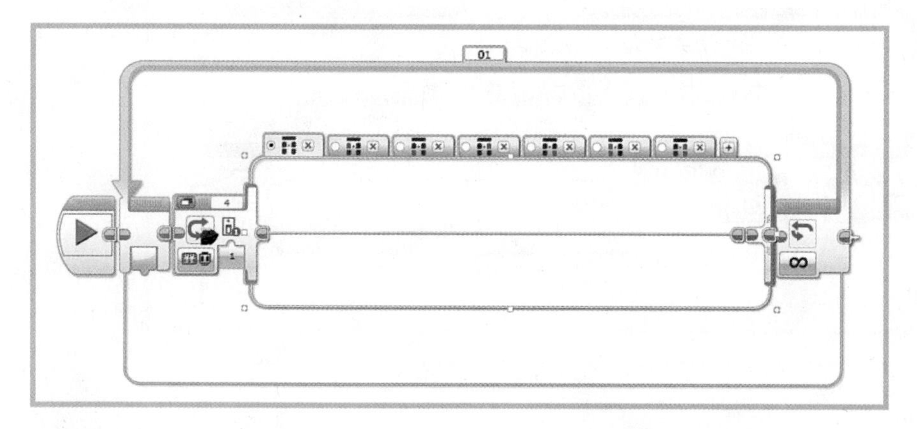

FIGURE 4.66

STEP 6 Next populate each tab with the appropriate motor block. Begin with the tab with the upper left button pressed. This controls Motor A, which raises and lowers the cord. Place the Medium Motor Block in the tab and leave it at its default settings (see Figure 4.67). Basically, when this program runs, you rotate the winch when you press that button.

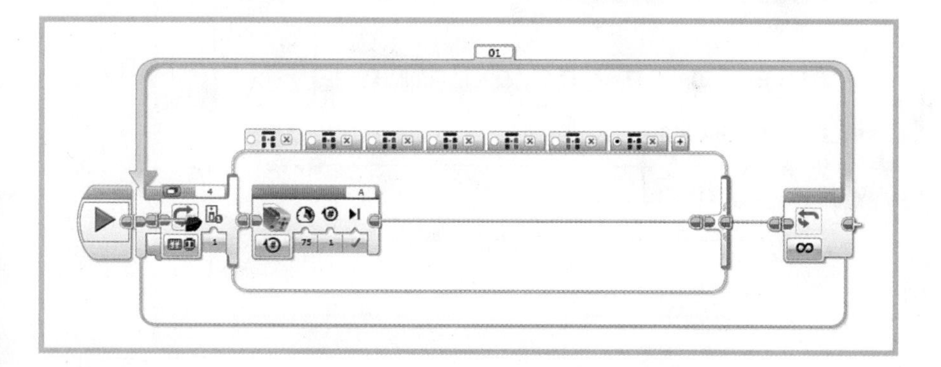

FIGURE 4.67

STEP 7 The next tab, you guessed it, merely reverses the previous one. The only change between the tabs, other than using the bottom button to connote reverse, is to change the rotation from 1 to -1. Figure 4.68 shows how it should look.

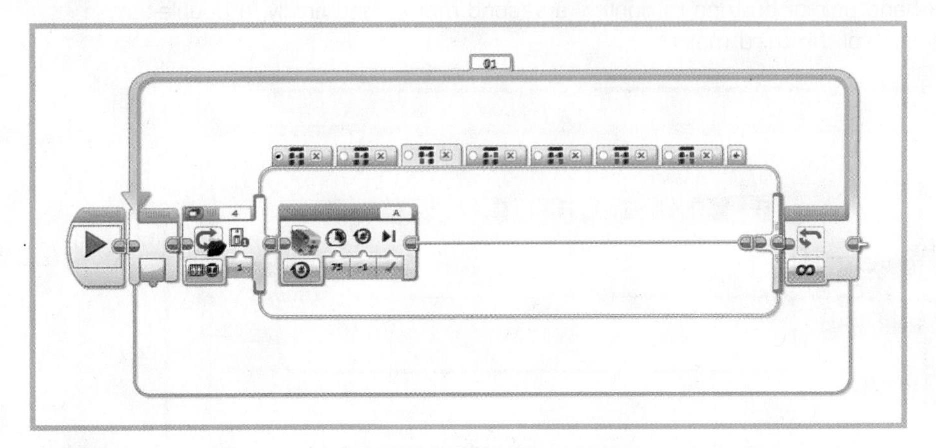

FIGURE 4.68

STEP 8 The next tab, which shows the upper right-hand button pressed, controls the rotation of the crane arm. Put a Large Motor Block in the tab and change its rotation to 0.1—you don't want the arm to move too far (see Figure 4.69)! Also, tone down its power to 25, so it doesn't move too quickly. This is Motor D in the program.

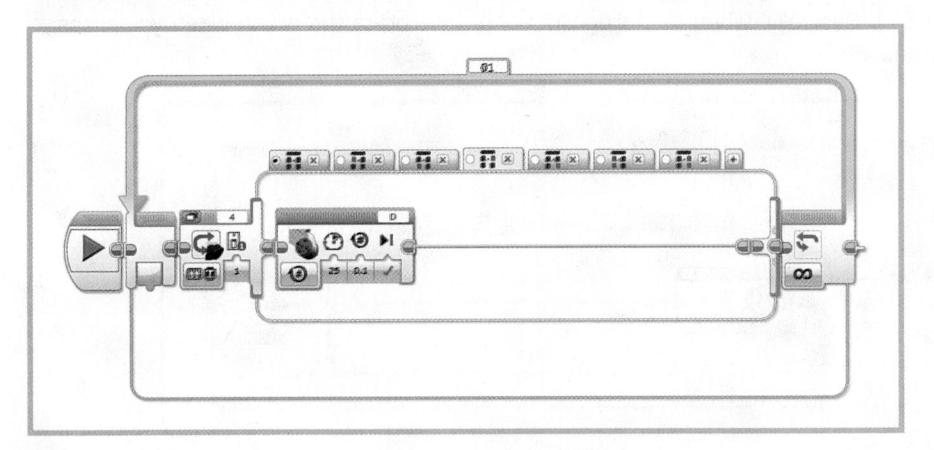

FIGURE 4.69

STEP 9 The tab marked with the lower right-hand button pressed does the expected and duplicates the previous tab except with a -0.1 rotation to move the arm the other way (see Figure 4.70).

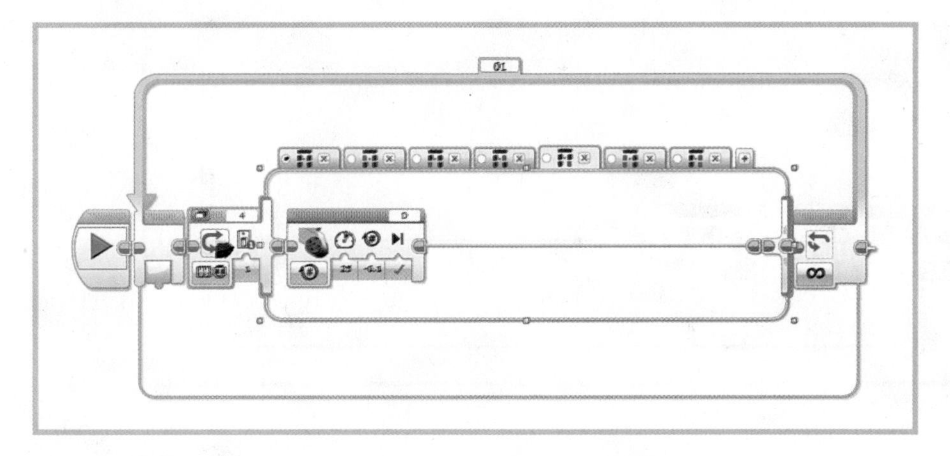

FIGURE 4.70

STEP 10 The last two tabs get a little tricky. You may have noticed that we're out of buttons. The solution is to designate these tabs with double-button presses (see Figure 4.71). Drag a Large Motor block into the tab, marking it as Motor C. This moves the robot forward on its wheels. The rotation can be changed to 0.25 and the power can be set to 25.

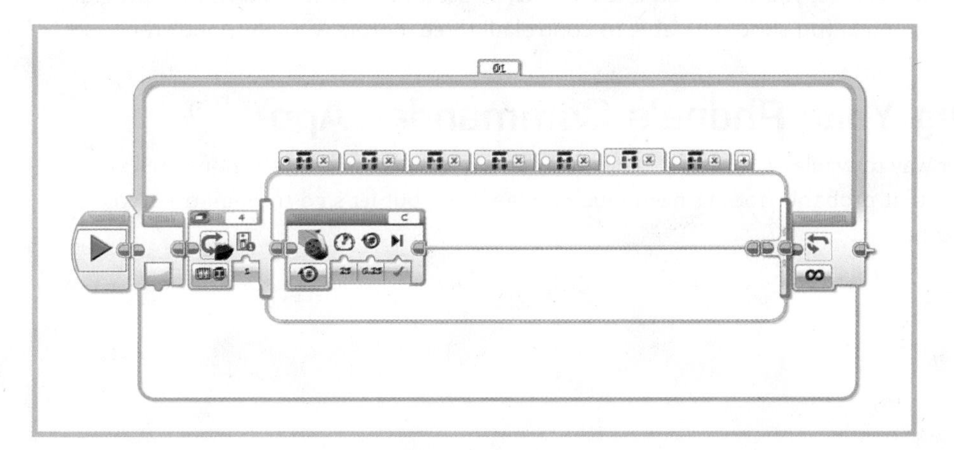

FIGURE 4.71

STEP 11 Finally, make the last tab a reverse of Step 10 as you'd expect. Figure 4.72 shows how it should look. You're done! Download and run the program and grab that remote.

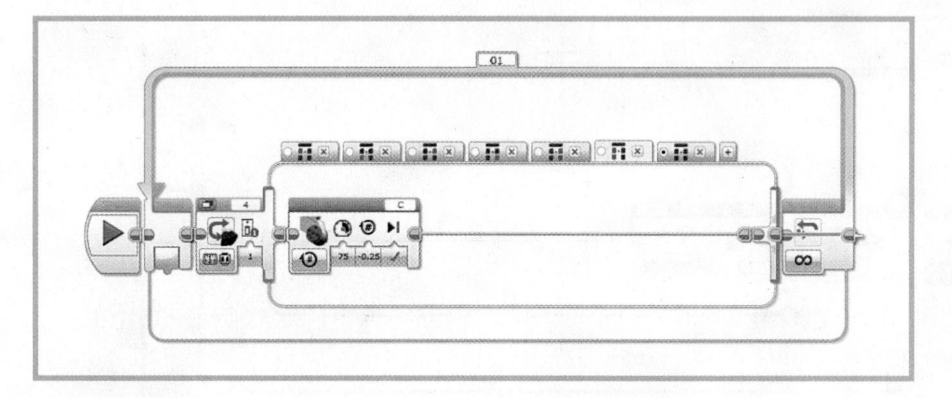

FIGURE 4.72

Controlling the Crane with Infrared

If you've looked at your EV3 remote control—excuse me, Infrared (IR) Beacon—you'll notice it has a red slider button with four settings. This is your channel selector. This lets you potentially control four different robots, or parts of a complicated one, with a single remote. In this case, you only need one channel, so set it to 1. When you run the program on the EV3 brick, you should be able to control all three motors with your button-presses.

Using Your Phone's Commander App

Another way to wirelessly control the Crane is through Bluetooth, a common wireless protocol that probably doesn't need much explanation, but let's go through the steps to get it set up.

STEP 1 You need a Bluetooth-equipped wireless device, operating on either Android or Apple's iOS operating systems. LEGO's Commander app may be freely downloaded, so go ahead and get it installed and then launch it (see Figure 4.73).

FIGURE 4.73

STEP 2 Click on the Bluetooth icon on the left-hand corner of the app and follow the directions to connect to the EV3 brick with your device (see Figure 4.74).

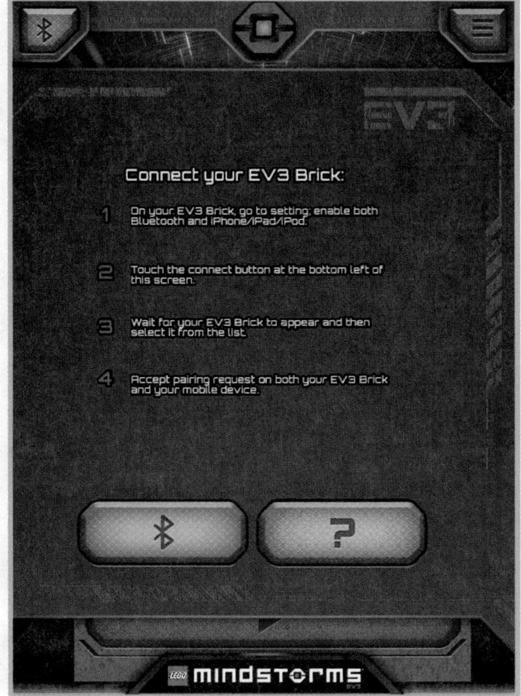

FIGURE 4.74

STEP 3 When you're connected, the screen shown in Figure 4.75 appears when you tap on the Bluetooth icon again.

FIGURE 4.75

STEP 4 We're ready to go. Tap on the right arrow button and cycle past the R3PTAR and the GRIPP3R robots—these are preconfigured control setups. Obviously, this is a nonofficial robot, so you would have to create your own interface. Past those two is Create & Command Your Own Robot (see Figure 4.76). That's the one you want!

FIGURE 4.76

STEP 5 The next page consists of a grid of 20 cells, as shown in Figure 4.77. Each of these is a possible component in your control interface (such as a joystick, button, or slider). Tap on one at random.

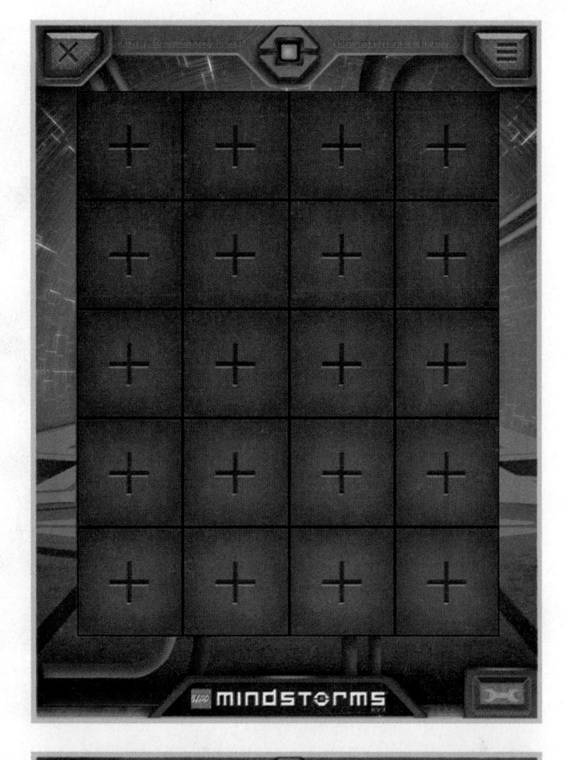

FIGURE 4.77

STEP 6 The first option of several is the Joystick (see Figure 4.78). Cycle through the various components by hitting the arrow buttons. A lot of the options look really cool!

FIGURE 4.78

STEP 7 Go to Horizontal Slider and select Motor C, as shown in Figure 4.79. Then add the slider to the matrix by hitting the + at the bottom of the screen.

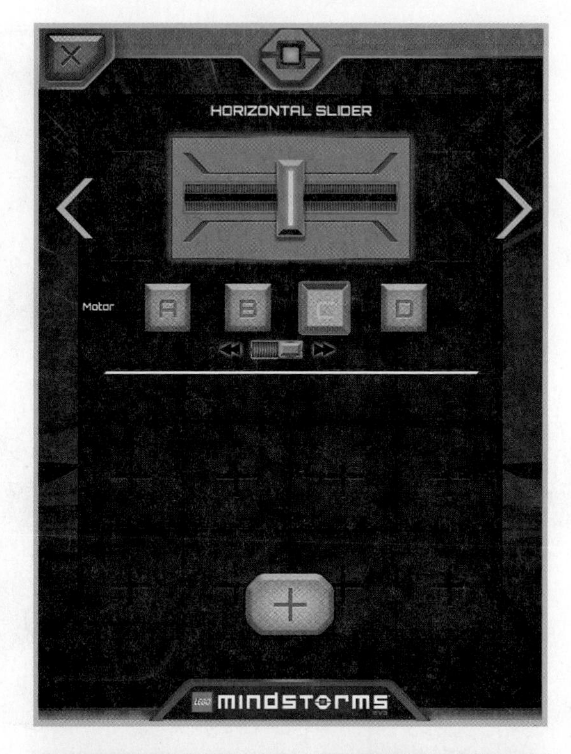

FIGURE 4.79

STEP 8 The translucent image of the slider appears on your screen, as shown in Figure 4.80. The reason you can see through it is because you're still editing the interface. Toggle in and out of edit mode by hitting the wrench icon on the bottom right-hand corner of the screen.

FIGURE 4.80

STEP 9 Next, add a joystick, set to Motors A and D. Because of space limitations in the interface, there really isn't room to have three sliders, so you control the winch and arm through the joystick. Figure 4.81 shows the matrix with the joystick added.

FIGURE 4.81

STEP 10 Click on the wrench to toggle out of edit mode. The icons turn opaque and you may now control your motors (see Figure 4.82)!

FIGURE 4.82

The app saves your interface for future use. You're golden!

Summary

After building the Remote Controlled Crane, you learned how to wirelessly control the robot using two different techniques: infrared remote control and a Bluetooth app on your smartphone. In Chapter 5, "Hacking LEGO II: Alternate Microcontrollers," you delve into non-LEGO ways of controlling a Mindstorms robot besides the EV3 Intelligent Brick.

5

Hacking LEGO II: Alternate Controllers

Sometimes LEGO hackers feel compelled to cast aside the (actually pretty sophisticated) EV3 Intelligent Brick and use some other control system on their robot. They choose microcontrollers such as the Arduino UNO and microcomputers such as the Raspberry Pi and BeagleBone Black to serve as the brains of their project (see Figure 5.1). It turns out there are plenty of cool reasons for swapping out the EV3. In this chapter, I go over some of the major factors in choosing whether to use one of these devices and which one. Finally, I share some cool no-EV3 projects I have encountered.

FIGURE 5.1 The Arduino UNO controls robots—even Mindstorms ones!

Microcontrollers and Microcomputers

Let's jump in feet first by exploring two cool projects that have been adapted to work with Mindstorms components. The Arduino is a microcontroller, essentially a complicated switch that triggers some pins when certain events occur. By contrast, a Raspberry Pi is a microcomputer—an actual computer on a chip, capable of doing (nearly) everything a full-sized computer can accomplish. Let's check out these two worlds.

Arduino

This small circuit board consists of an ATmega328P microcontroller chip, with its input and output functionality managed by a series of connectors called pins. The Arduino runs simple programs that take input from sensors and use that data to control motors, displays, and other components.

Running off a 9V battery and measuring only about 3 inches across, the Arduino is ideal for small robotics projects. It uses its 14 digital and 6 analog pins to interface with other components and devices, giving it the capability to control an impressively complicated project.

If you're looking to buy an Arduino, you can find them all over, but my favorite online store, SparkFun Electronics (sparkfun.com) sells them, along with countless other electronic and mechanical accessories.

Let's go over the various features of the Arduino UNO, following along with Figure 5.2:

1. **Reset button**—This button starts the Arduino's program over from the beginning.
2. **Digital pins**—These pins receive or send data that is either a 1 or 0, or a whole bunch of 1s and 0s: This is called digital data, contrasting with analog data, which is a continuous stream of values.
3. **Circuit board**—This printed circuit board (PCB) is embedded with traces (AKA wires) that connect the various components together.
4. **USB jack**—This standard USB-B female jack not only can power the Arduino but also controls data input.
5. **Power indicator LED**—Stays lit as long as the Arduino is powered on.
6. **Data input/output LEDs**—These LEDs light up when data transmission takes place.
7. **ATmega328P microcontroller**—The brains of the operation.
8. **Power jack accepting a 2.1mm barrel plug**—This allows you to run your Arduino from house current—no more wasting batteries!
9. **Power management pins**—These pins provide 3.3-volt and 5V sources for your projects.
10. **Analog pins**—The counterpart of the digital pins, the analog pins are great for taking analog sensor readings and for handling potentiometers.

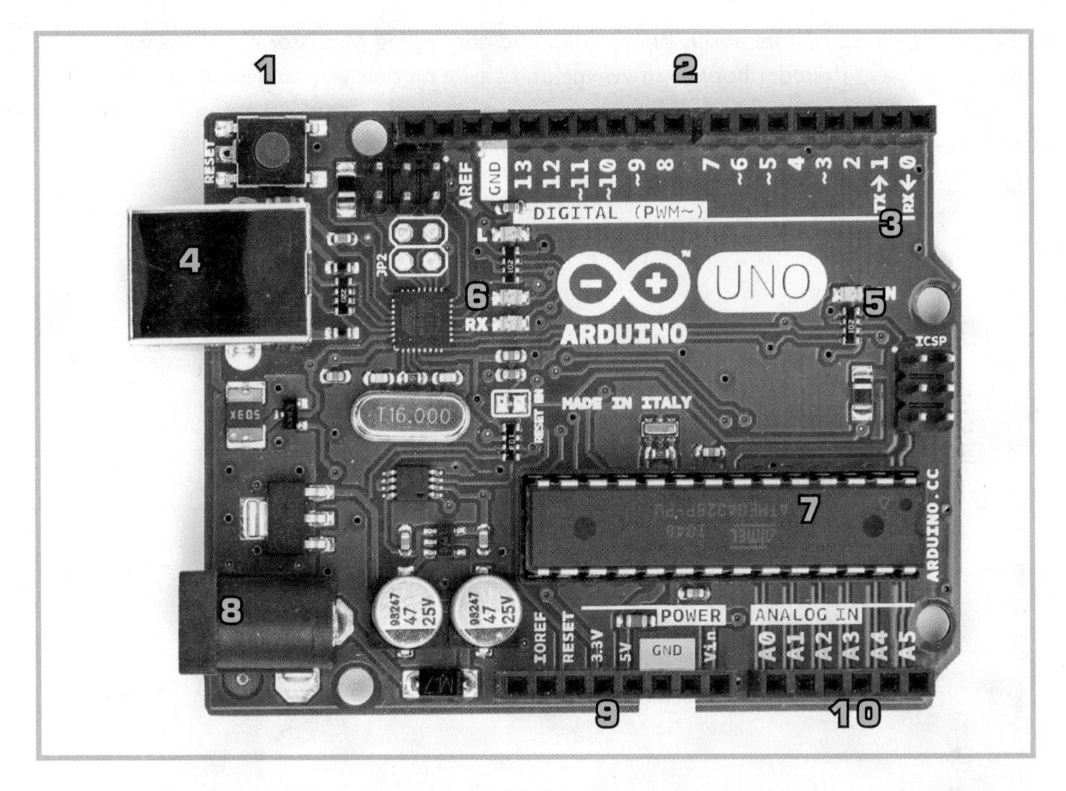

FIGURE 5.2 The Arduino UNO features 14 digital and 6 analog input/output pins.

So that's an Arduino, but how does that let you control a Mindstorms robot? It doesn't even have the right kind of sockets! Let me introduce you to the Bricktronics shield, shown in Figure 5.3. The shield fits on top of the Arduino and interfaces it with. It can control two motors (to the EV3's four), four sensors, and even has a couple of transistors enabling you to trigger non-Mindstorms components such as DC motors.

One of the first forays into Arduino control of Mindstorms, Bricktronics is the creation of Minneapolis-based kit makers Wayne and Layne (wayneandlayne.com). They wanted to interface their Arduino with Mindstorms motors and sensors but were stymied by the proprietary Mindstorms plugs with the off-center tab (mentioned in Chapter 2, "Project: Plotter Bot"), so they had a crate of sockets manufactured overseas, giving them a huge supply for their own projects.

Wayne and Layne also wrote an Arduino library for Bricktronics. Libraries are code resources that allow you to control complicated components without bogging down your sketch (as Arduino programs are called) with huge amounts of uber-complicated code. The complicated stuff is held in a different file, the library, so your sketch remains nice and clean. The Bricktronics library includes code examples for a variety of projects, illustrating a number of techniques for controlling Mindstorms motors and sensors. You can download the library from Wayne and Layne's Bricktronics site, http://www.wayneandlayne.com/bricktronics/.

If you want to learn more about Bricktronics and controlling Mindstorms robots with Arduinos, I coauthored a book with Wayne and Layne on that very subject. Called *Make: LEGO and Arduino Projects* (Make 2012), the book explains how to build a number of models and covers electronics, programming, and Arduino lore.

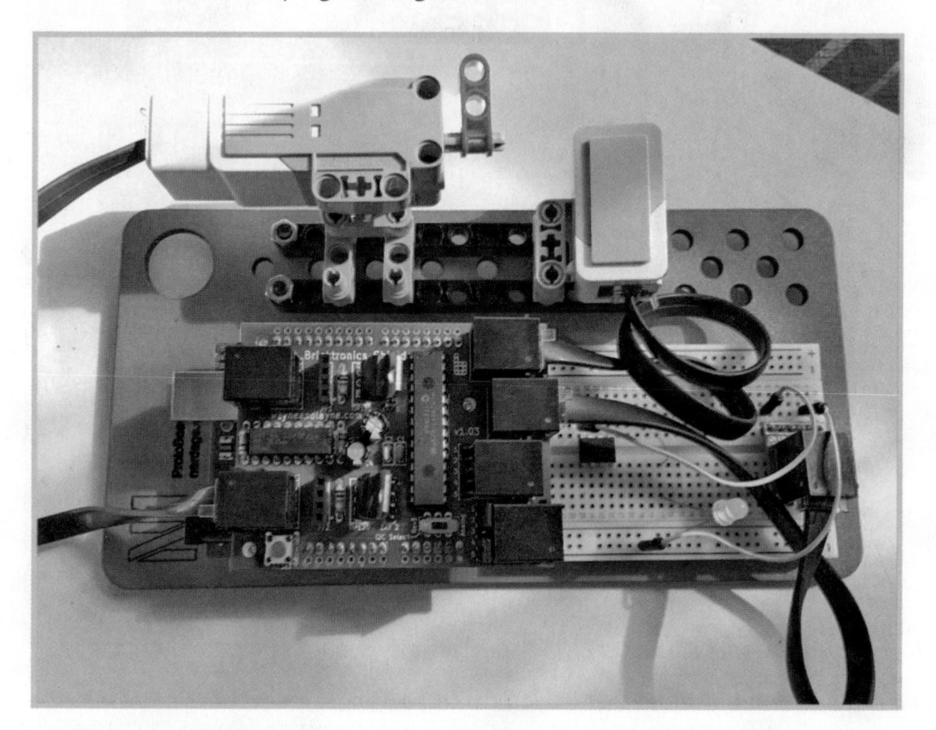

FIGURE 5.3 The Bricktronics shield fits on top of the Arduino.

Raspberry Pi

Much more sophisticated and complicated than an Arduino, the Raspberry Pi has much more capability but also sports a higher learning curve. Still, the microcomputer has found its niche, with 5 million Pi's sold since the board debuted March 14, 2012.

Raspberry Pi's are credit-card-sized computers, equipped with USB, RCA, and HDMI ports so you can plug one into a TV and add a keyboard and mouse to make it actually work as a computer—granted, not an impressive one. You can still surf the Web and check email on it, and even run simple applications such as a word processor. All that, and it works like an Arduino with input-output pins awaiting your favorite add-on modules, sensors, and motors.

The Pi features one key improvement over both the EV3 and the Arduino UNO—it features built-in Internet connectivity through its Ethernet socket, but also through add-on Wi-Fi and wireless modules. You can buy specialized add-on boards and modules that add these capabilities to an UNO, but the Pi does it natively.

If you want to purchase a Pi, you can find the latest version, the B+ (P/N 12994) at SparkFun.

Let's go over the Pi and its various features, following along with Figure 5.4.

1. **Input/output pins**—The B+ sports 40 GPIO (General Purpose Input Output) pins, some of which are for power and ground, but still an impressive number considering that many sensors and add-on modules need only one data pin to work.
2. **USB**—The B+ boasts four USB 2.0 sockets.
3. **Display**—This display connector is meant to accommodate an LCD screen for interfacing with the Pi.
4. **SD card**—Beneath the display connector you find a micro SD card slot. This is the memory storage of the Pi, including its operating system.
5. **CPU**—The brains of the board. As mentioned, it's an actual computer-on-a-chip capable of running Linux.
6. **Camera**—Plug certain camera modules into this port.
7. **Network controller**—This chip helps interface the Pi with the local network via the Ethernet socket.
8. **Micro USB**—The Pi may be powered through this port, and the software is updated through this cable.
9. **HDMI**—A standard HDMI port that allows you to plug the Pi into a TV.
10. **Audio**—Stereo audio and composite video combined into one.
11. **Ethernet**—Plug your Pi into the network here.

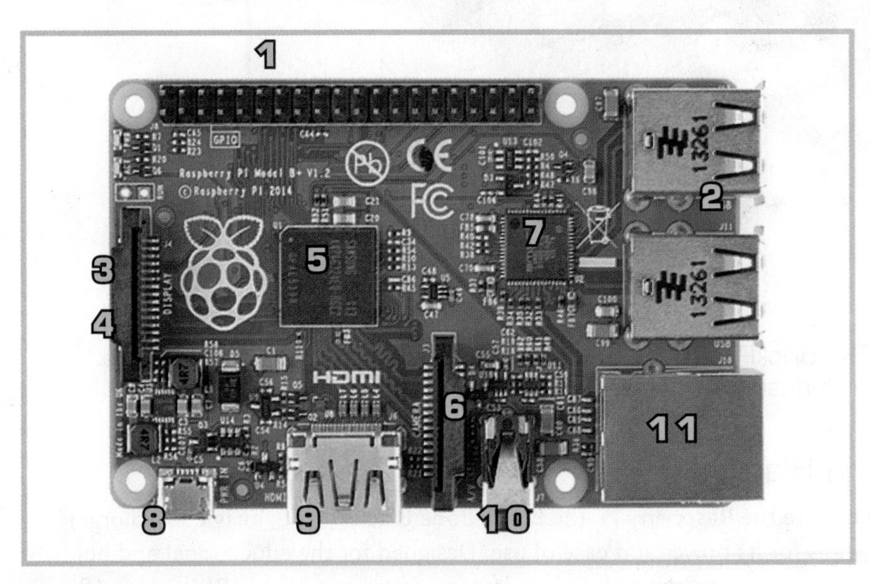

FIGURE 5.4 The Pi looks a little like an Arduino but packs much more of a punch.

Just as the Arduino has the Bricktronics interface board, Raspberry Pi has the BrickPi, a creation of Dexter Industries. It's an interface board that plugs into the top of the Pi, allowing you to control motors and take readings from sensors (see Figure 5.5).

The BrickPi can control up to five Mindstorms sensors and four motors, making it a worthy competitor to the EV3 brick, and the Ethernet/wireless capabilities of the Pi suggest that some crazily complicated inventions can be built. On that note, I share a couple of these excellent projects later in this chapter.

FIGURE 5.5 The BrickPi allows you to control five sensors and four motors. Credit: Dexter Industries.

BeagleBone Black

A microcomputer like the Raspberry Pi, the BeagleBone Black (http://beagleboard.org/) has won many fans for its power and ease of use. Designed for the educational and hobby markets by Texas Instruments, the BBB (as it's known) features a 1Ghz ARM-Cortex A8 processor, half a gigabytes of RAM and 2 gigabyte on-board storage—respectable computer stats, in other words. Like the Raspberry Pi B+, the BBB sports Ethernet, microSD, HDMI, and USB ports. It features a whopping 92 header pins for maximum input and output, though some of these are reserved for LCD data.

Let's go over the various parts of the BBB; refer to Figure 5.6 as you read down the list.

1. **46-pin headers**—These two rows offer considerable resources for interfacing with a complicated robot.
2. **DC barrel plug**—This plug allows the BBB and its project to be powered by a wall wart.
3. **Ethernet socket**—Like the Raspberry Pi, the BBB can be plugged directly into the network.
4. **The ARM-Cortex A8 processor**—The brains of the board.
5. **RAM**—The BBB includes 512MB in DDR3 RAM.
6. **Two USB ports**—Connect a keyboard or mouse to these ports.
7. **HDMI**—The BBB has a micro HDMI socket on the underside.
8. **Storage**—This chip gives the BBB 2 gigs of on-board storage, plus whatever the SD card holds.

The platform has been around for a few years, and a variety of add-on boards called *capes* have been created, much the way Arduino shields add on sensors and displays to the main board. You can buy BBBs and capes (add-on boards) at SparkFun, (P/N 12076 for the BBB).

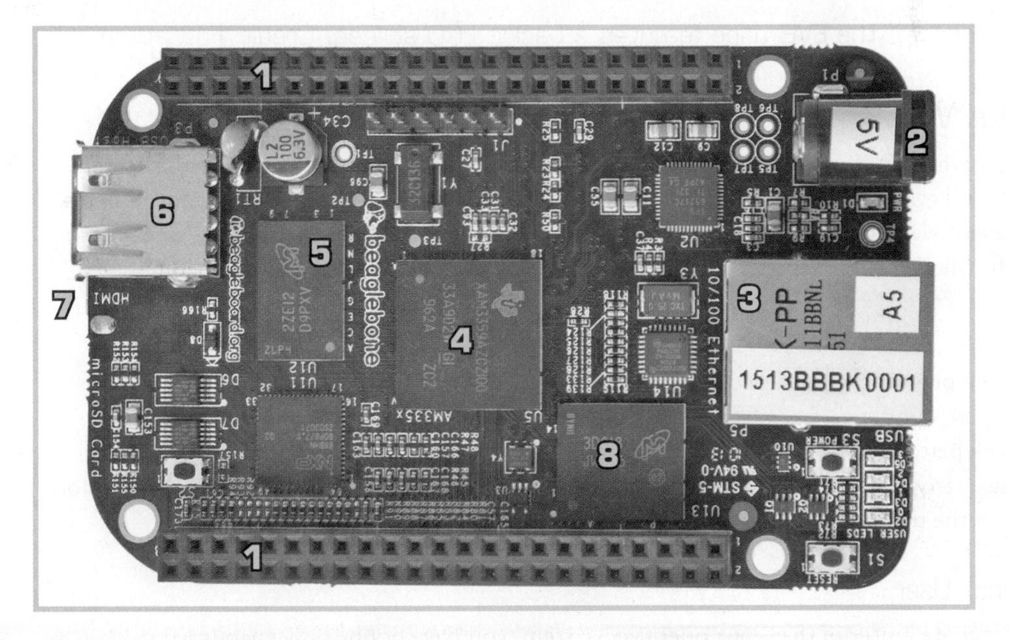

FIGURE 5.6 The BeagleBone Black, like the Pi, features a full-fledged computer on a chip.

There is a Mindstorms-controlling cape for the BBB, so new that, as I write this in April 2015, its crowdfunding campaign is still in full swing. Called EVB (a "BBB" twist on EV3), it features eight sockets for Mindstorms wires, and as shown in Figure 5.7, it includes a backlit color LCD screen as well. If you're interested in learning more about this new technology, check out the EVB's home at http://www.fatcatlab.com/.

FIGURE 5.7 The EVB cape features a backlit LCD screen. Credit: FatCatLab.

But Why?

I get what you're saying. The EV3 Intelligent Brick is a perfectly robust and sophisticated controller. By contrast, an Arduino UNO is relatively limited in certain important respects. Nevertheless folks quixotically continue to explore these alternatives. Here are some reasons why Mindstorms hackers like to swap in different microcontrollers and microcomputers.

Preference

Some people flat-out don't want to learn a new system. This may seem like an arbitrary reason, but it's perfectly legitimate. The best reason to hack something is to make it work better for you, and if that means adapting all your Mindstorms robots to work on Raspberry Pi boards, so be it! After a certain point, you just want things to work—you don't have the time to learn a new interface.

Linux User

If you are a user of the Linux operating system, you have doubtlessly encountered numerous situations where a particular application didn't support your OS. The Mindstorms software falls into this category. If those worthy folks want to make

Mindstorms robots, they are forced to use emulators or use a different computer. However, alternate EV3 firmware such as LeJos (described at the end of the chapter) as well as the Arduino, Raspberry Pi, and BeagleBone Black all are Linux compatible, making it a no-brainer to move away from the default.

Education

Engineering teachers love LEGO Mindstorms, but as they teach it, they encounter the truism that they have to jump to a new building system or kit when the students outgrow Mindstorms' rather limited programming interface. One school of thought is to reflash the Intelligent Brick with another operating system such as Java or C. I go into this topic at the end of this chapter, so stay tuned!

Another approach, as I already mentioned, involves swapping in electronics suites such as Arduino/Bricktronics and RaspberryPi/BrickPi. The robot the students built remains the same, while only the control system changes. Helping students perform the same task in two different systems offers a great learning opportunity.

These devices represent "step two" in teaching programming and robotics. They also teach electronics, if only because they are not sealed up in a plastic box, so you can learn about what each part does.

Power

The Arduino can't compete with the EV3 brick, but the Raspberry Pi and BeagleBone Black are actually miniature computers, as I already mentioned.

They're so robust you can stream movies or surf the Internet with them. To a degree, the EV3 Intelligent Brick has a similar level of computational power, but it's hindered by the controlled and locked down interface. The EV3 was designed for middle-school students or thereabouts, and limiting capabilities in exchange for a simpler interface seemed (and probably is) a good idea.

That said, if you're up for a challenge and willing to learn Python scripting (if you don't already know it), then you can get some startling power out of a Pi or BBB.

Ecosystem

The open source hardware and software communities are vast, at least compared with adult Mindstorms developers. This means many more brains to put together and many avenues being explored.

It's hard to match the ecosystem—the sheer breadth of electronic sensors, integrated circuits, and other modules that fans have gotten to work with Arduinos, for instance. Take the Chronodot, a real-time clock module that keeps nearly perfect time, more accurately than most microcontrollers (see Figure 5.8). If you want to use it with an Arduino or Pi, no worries. Good luck trying to get it to work with your EV3 brick.

It's not just hardware. There is an ecosystem of code as well, vast collections of programs and libraries exploring every major component these hackers can get their hands on. One site exploring this code, the Arduino Playground, serves as a huge sandbox to learn about pretty much everything relating to Arduino software (http://playground.arduino.cc/).

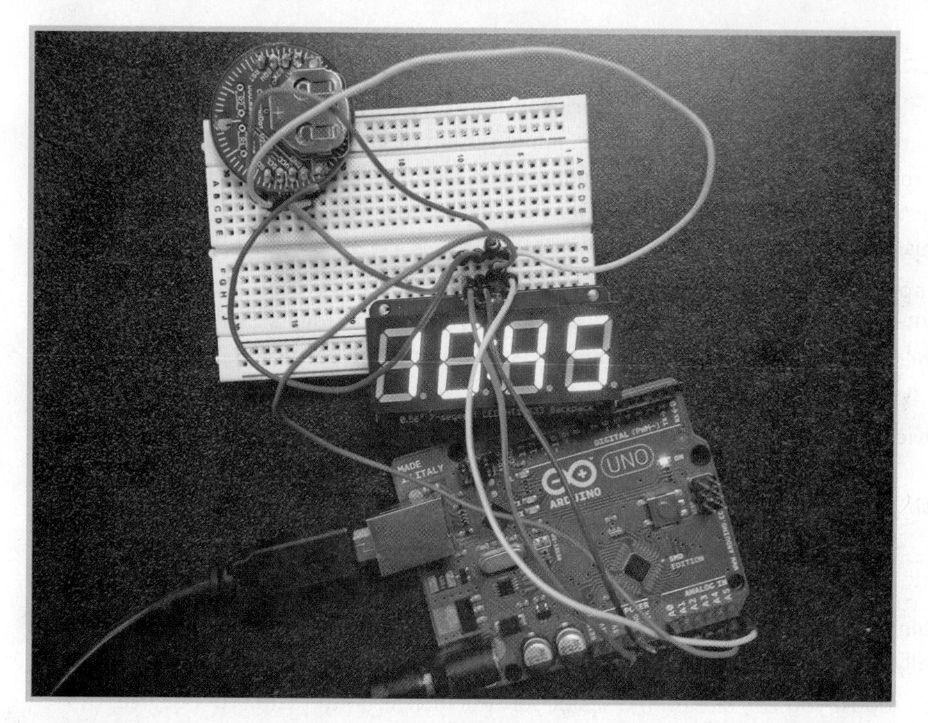

FIGURE 5.8 The Chronodot (the round circuit board) helps the Arduino keep nearly perfect time.

Example Projects

Here are a couple of Mindstorms projects that show a range of what can be done with an Arduino or Raspberry Pi.

Book Reader

If you have an old book and want to digitize it, this robot scans the text with an add-on camera module and then converts the images to text, allowing the robot to convert printed materials into data (see Figure 5.9).

Featuring a Raspberry Pi and BrickPi board, the Pi makes use of a powerful but inexpensive camera module that plugs right into the circuit board—just try doing that with your EV3 brick!

http://makezine.com/projects/lego-bookreader-digitize-books-with-mindstorms-and-raspberry-pi/

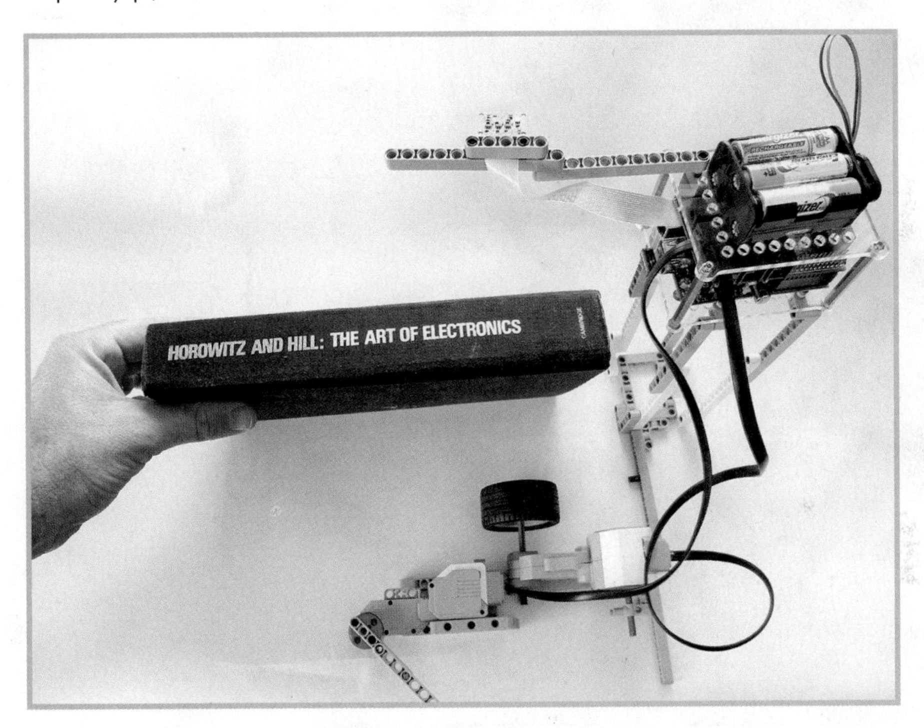

FIGURE 5.9 The book reader digitizes printed books. Credit: Dexter Industries.

Mini Tank

Designed to be the basic starter robot, the Mini Tank can be built from a single Mindstorms kit and accommodates all the usual sensors and motors (see Figure 5.10). The top features a flat mounting point where an Arduino, Bricktronics shield, and battery pack can be mounted. I show how to build the Mini Tank in my book *Robot Builder: The Beginner's Guide to Building Robots* (Que 2013).

FIGURE 5.10 The mini tank is designed to be the basic Bricktronics robot.

Chocolate Milk Maker

The Chocolate Milk Maker was a well-intentioned drinkbot: a robot that prepares beverages (see Figure 5.11). This one was supposed to make chocolate milk, but dispensing syrup and milk, stirring the mixture, and keeping everything clean turned out to be just as much of a challenge as getting the mix right. I built this project along with Wayne and Layne as part of our *Make: LEGO and Arduino Projects* book I mentioned. If you want to build your own, check out our book!

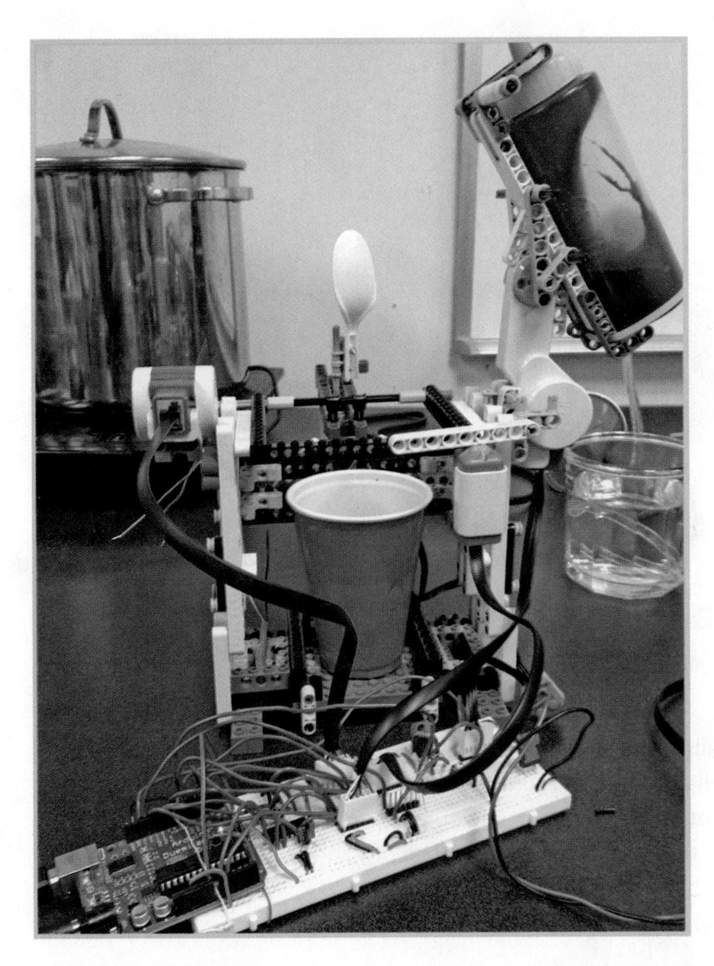

FIGURE 5.11 The Chocolate Milk Maker mixes up a glass of chocolate milk.

Ball Counting Robot

This robot handles ping-pong-sized balls (see Figure 5.12). It's based off a Ball Contraption, a category of LEGO robot that moves a handful of balls along a mechanical circuit, either transporting the balls to a new module or cycling them through repeatedly. What the Ball Counting Robot does is...well, it counts the balls as they pass by.

This robot uses a proprietary Arduino-based microcontroller with RJ25 cables providing connectivity. You're probably asking yourself, how do they connect the RJ25s to the Mindstorms components? The answer is that they neatly sidestepped the issue by using non-LEGO electronics.

http://makezine.com/projects/build-a-ball-counting-robot-using-makeblock-and-lego/

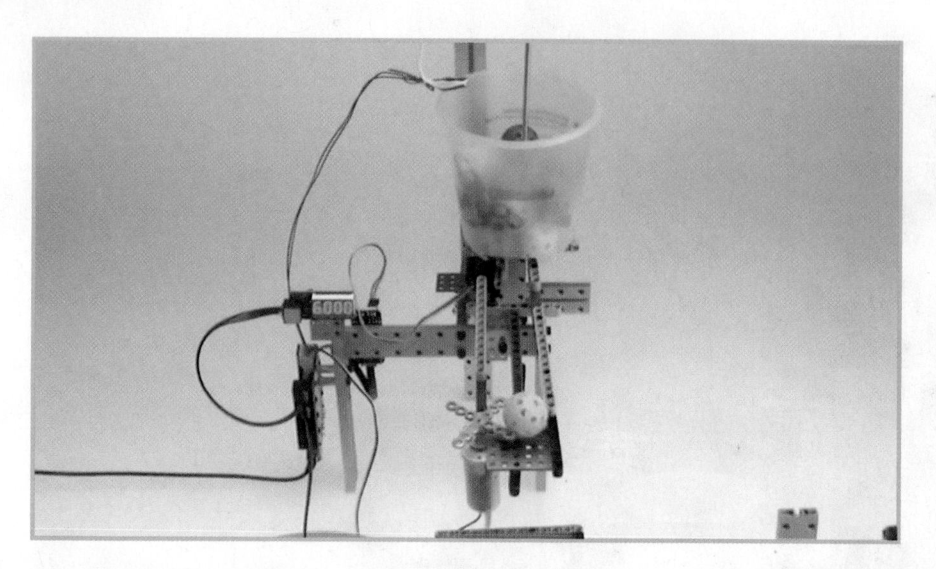

FIGURE 5.12 The Ball Counting Robot keeps track of the number of balls cycling through the contraption. Credit: Makeblock.

Rolling Alarm Robot

The creators of the BrickPi built the Rolling Alarm Robot with the Raspberry Pi, and the annoying creation rings its bell and dodges around, forcing you to wake up fully to turn it off (see Figure 5.13). The robot takes its cues from your own Google Calendar account, so you can schedule your waking times from the convenience of your nearest computer, tablet, or phone.

http://www.instructables.com/id/Rolling-Alarm-Robot/

FIGURE 5.13 The Rolling Alarm Robot makes you chase it. Credit: Dexter Industries.

BeagleBone Black Robot

The creators of the EVB interface board (mentioned earlier in the chapter) assembled a Gyro Boy, a frequently built type of robot that balances on two wheels, using a Mindstorms gyro sensor to detect when it's tipping over and moving the wheels to correct its balance (see Figure 5.14)—kind of like a Segway, only little and cute. Then they removed the EV3 brick and added a BeagleBone Black and EVB cape—sort of a fun and simple robot to help newbies get their feet wet.

http://www.instructables.com/id/EVB-A-way-to-replace-the-brain-of-the-LEGO-Mindsto/

FIGURE 5.14

The Gyro Boy gets a brain transplant. Credit: FatCatLab.

Programing Environments

Let's close out the chapter by switching gears a bit. What happens when instead of swapping out the EV3, you simply reprogram it? It turns out that other operating systems can be loaded onto the EV3 Intelligent Brick, completely eliminating Mindstorms' graphical interface and replacing it with a professional programming environment. The book isn't focusing on this technology, but here's a list of resources if you should go down this path.

LeJos

This replacement firmware lets you run a Java Virtual Machine, which allows you to write programs in the Java programming language. As such, it has advanced programming features one would associate with a language used by professionals.

An open-source project around since 1999, LeJos has a wealth of resources, such as example code available to it. Because of this, and its connection to LEGO robotics, LeJos is often used to teach programming to engineering students.

http://www.lejos.org/

RobotC

RobotC replaces the firmware with a C programming environment. It's a great way to teach a high-level language to older kids who may no longer find Mindstorms challenging.

RobotC costs money, which excludes some folks, but it's aiming at the high-powered realm of high school robotics championships. Participants want a polished and sophisticated platform and are willing to pay money to get it.

Cleverly, the RobotC folks also provide versions of the firmware for the VEX controller module as well as Mindstorms, giving them access to two classes of educational robots. However, unfortunately, they offer their software only for Windows-based PCs, so Mac and Linux users will have to use an emulator.

http://robotc.net

Monobrick

A firmware replacement for the EV3 brick, Monobrick uses Mono, an open source .Net implementation. You can program it in C#, F#, and IronPython, and unlike other firmware replacements, you can boot Monobrick right off the SD card. Compilers are available for Windows, Mac, and Linux.

http://www.monobrick.dk/software/ev3firmware/

ev3dev

A lot of alternate operating systems and firmware replacements assume you're wiping the Intelligent Brick and replacing its software with something new. ev3dev takes a different tack. You put the software on a SD card, and it runs Debian Linux on the Intelligent Brick's ARM9 processor. Using Linux, an open-source operating system, gives Mindstorms hackers access to a vast array of scripting languages, hardware options, and so on.

http://ev3dev.org

Summary

If you weren't sold on alternate controllers for Mindstorms robots before you read this chapter, I hope you are now! You got the straight scoop on three different powerful and robust platforms that can be swapped in. Now that you've familiarized yourself with the technology, it's time to put it to the test. In Chapter 6, "Project: Robot Flower," you build a robotic flower with your EV3 set and then you modify it to be controlled by an Arduino.

6

Project: Robot Flower

In this chapter you build another robot. This one is a Robot Flower that opens and closes its petals with the help of a pair of motors (see Figure 6.1). I show you how to build the Flower and then show you two ways to control it. The first method uses the usual EV3 Intelligent Brick, with the program moving the motors incrementally as commanded by the program. The petals begin the day tightly closed and then gradually open up as the day progresses until they are wide open. The process then reverses so the flower closes up as the day wanes.

After the initial build, I show you how to use an Arduino microcontroller in place of the EV3 brick, opening up a whole new world of robot control techniques.

FIGURE 6.1 The Robot Flower opens its petals as the day progresses.

Robot Flower Mindstorms Build

Let's begin the project by building the robot with all Mindstorms components. We'll worry about the Arduino later!

Parts List

You need the following Mindstorms parts to build the Robot Flower, following along with Figure 6.2. As usual, you can find these in your Mindstorms set.

- **A.** EV3 Intelligent Brick
- **B.** 2X large motors
- **C.** 2X 13M beams
- **D.** 4X 11M beams
- **E.** 4X 9M beams
- **F.** 6X 7M beams
- **G.** 6X 5M beams
- **H.** 4X 3x7 angle beams
- **I.** 2X 2x4 90-degree angle beams
- **J.** 4X T-beams
- **K.** 2X 5x11 chassis bricks
- **L.** 1X 5x7 chassis brick
- **M.** 3X 8m axles with end stop
- **N.** 1X 7M axle
- **O.** 4X 5M axles
- **P.** 2X 4m axles with end stop
- **Q.** 4X 3m axles
- **R.** 2X 3m axles with knob
- **S.** 2X 2m axles
- **T.** 5X bushings
- **U.** 2X half bushings
- **V.** 8X 3m pegs
- **W.** 55X connector pegs
- **X.** 6X connector pegs without friction tabs
- **Y.** 11X cross connectors
- **Z.** 4X cross block fork
- **AA.** 4X double cross blocks
- **BB.** 2X cross blocks
- **CC.** 2X 90-degree cross blocks
- **DD.** 1X 180-degree angle element
- **EE.** 1X 3m beam with pegs
- **FF.** 2X Z20 gears
- **GG.** 2X Z12 gears
- **HH.** 2X each left and right 3x7 and 3x11 panels

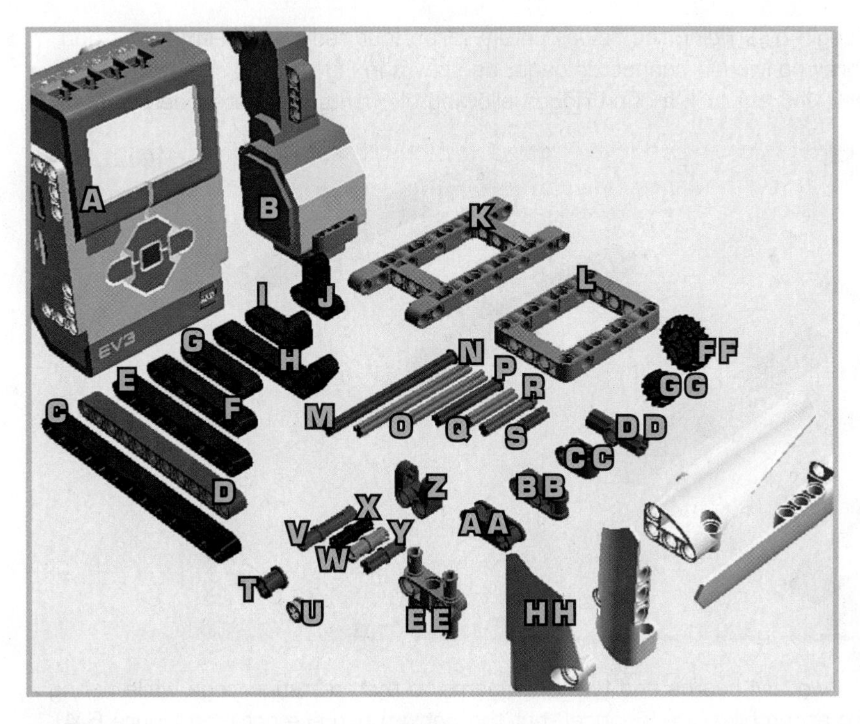

FIGURE 6.2 You need these parts to build the Robot Flower.

Steps

Once you have gathered together the Mindstorms parts—or simply have your bin handy!—you can start building the flower.

STEP 1 Let's begin by building the flower petals. Grab four red 11M beams, and add 3M pegs and gray no-friction connector pegs, as shown in Figure 6.3. They're just like the regular black kind but lack friction ridges allowing elements to rotate freely.

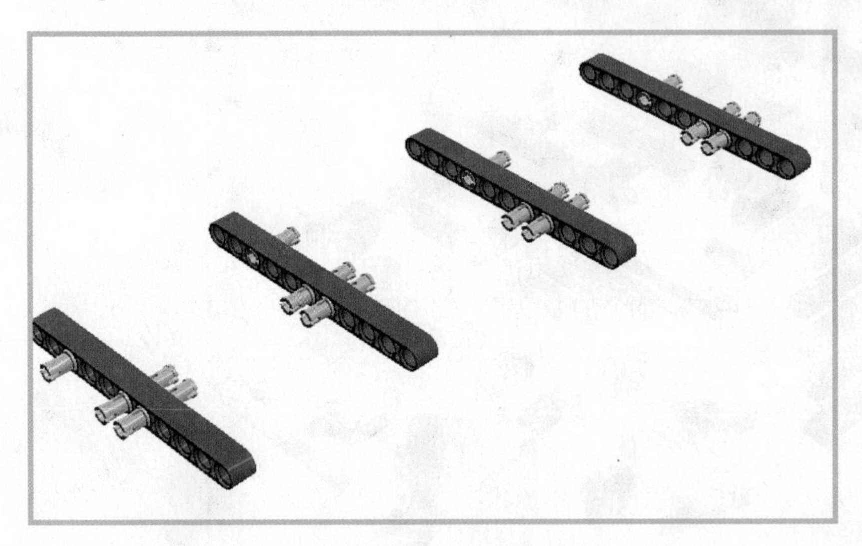

FIGURE 6.3

STEP 2 Attach two 5M beams and two 9M beams to the no-friction pegs while noting that the top two assemblies are identical, but the bottom two are not (see Figure 6.4).

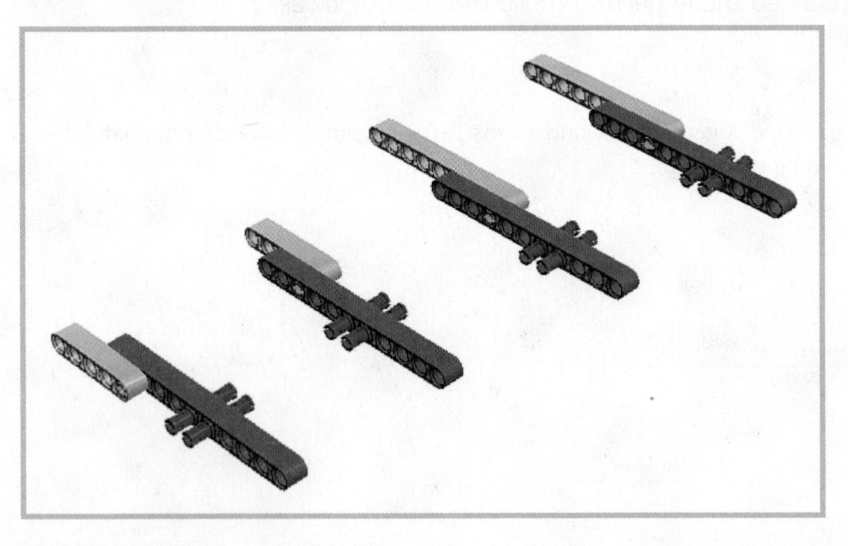

FIGURE 6.4

STEP 3 Insert two more of those nifty gray connector pegs. Your assemblies should look like Figure 6.5.

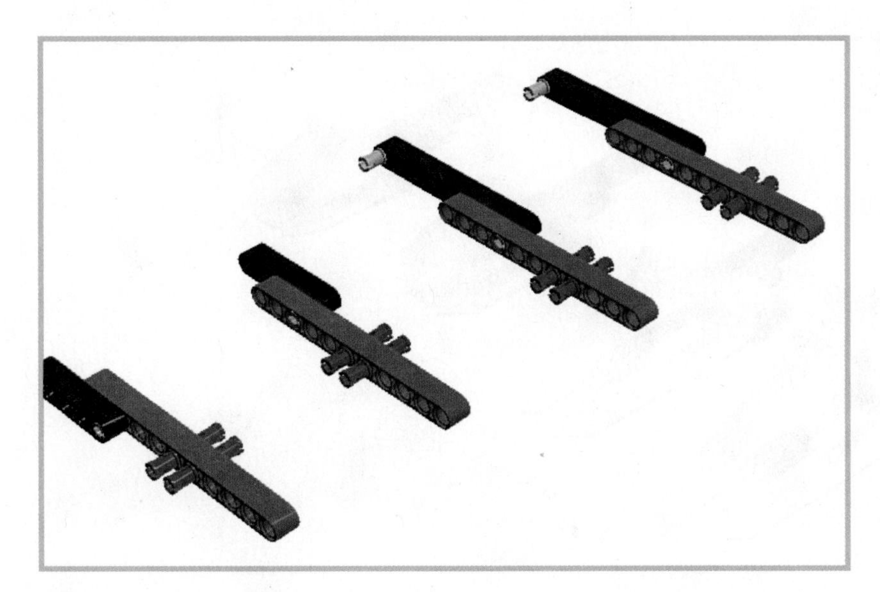

FIGURE 6.5

STEP 4 Next, add the 3x11 panels (as they are generically called) to the 3M connector pegs, as shown in Figure 6.6.

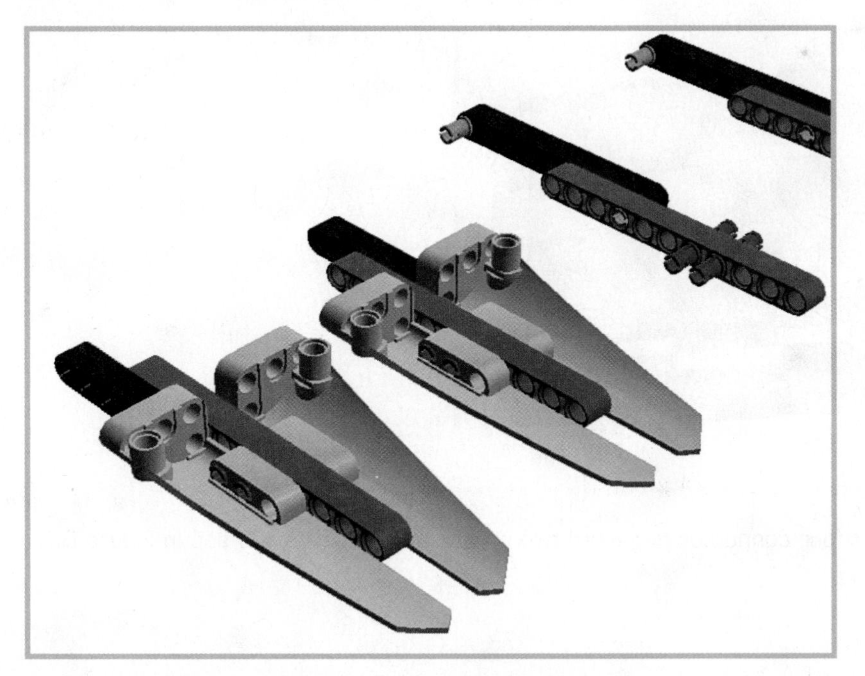

FIGURE 6.6

STEP 5 Attach the 3x7 panels to the other pair of assemblies (see Figure 6.7). Set aside the petals for now—we'll focus on the main build for now and then add the petals at the end.

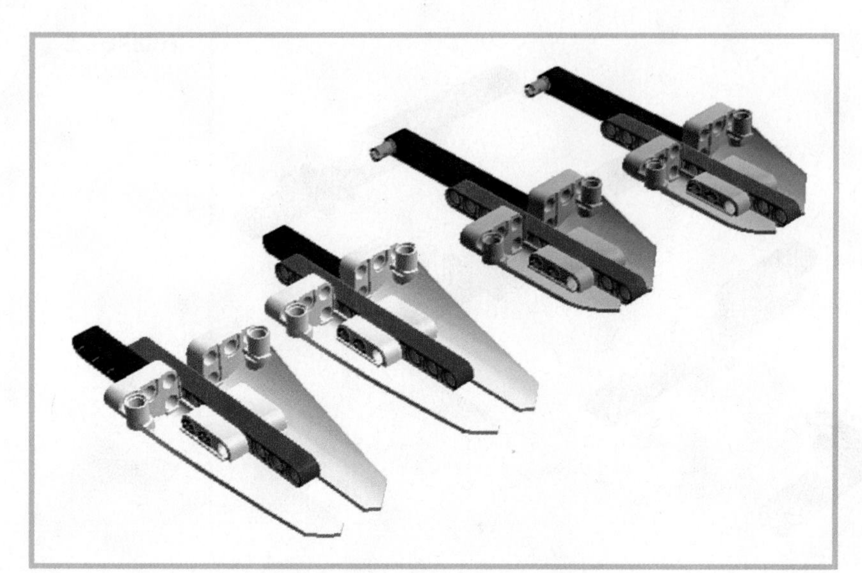

FIGURE 6.7

STEP 6 Attach a pair of 7M beams to the bottom of the EV3 brick, as shown in Figure 6.8. Use four connector pegs to secure them.

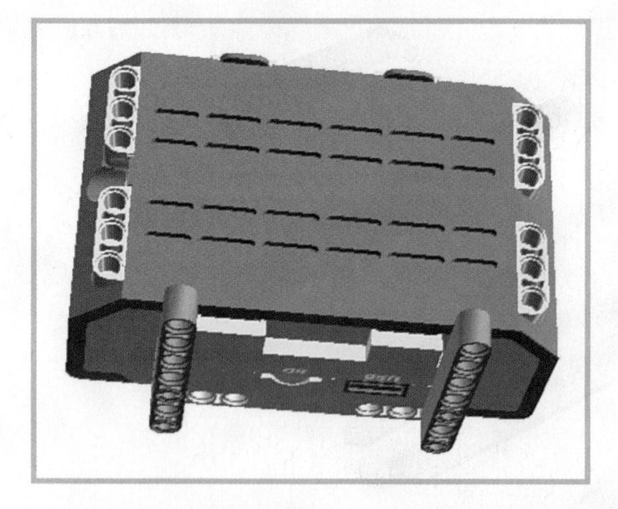

FIGURE 6.8

STEP 7 Insert four connector pegs and two cross connectors as you see in Figure 6.9.

FIGURE 6.9

STEP 8 Connect a pair of large motors to the pegs you just added, as shown in Figure 6.10.

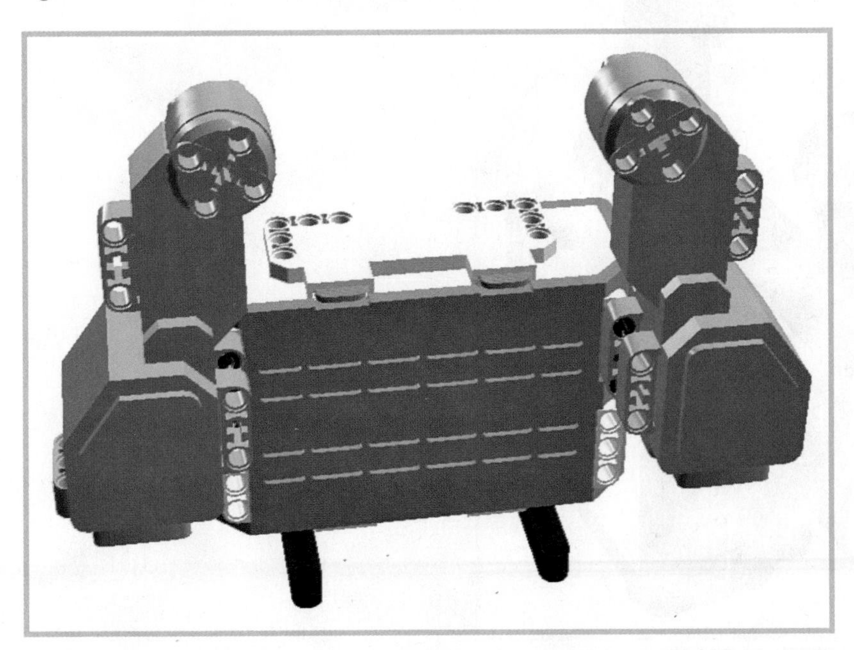

FIGURE 6.10

STEP 9 Secure the motors with a pair of 13M beams, each part held in position by a connector peg at either end. Figure 6.11 shows how it should look.

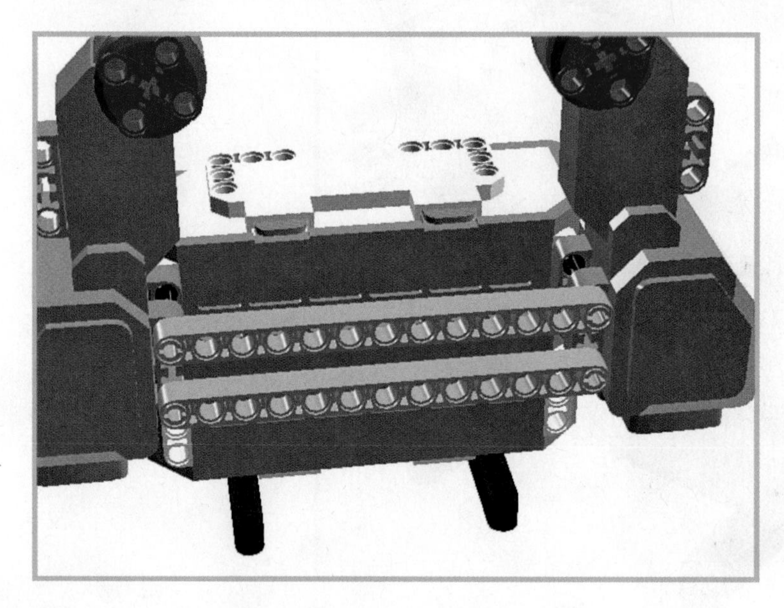

FIGURE 6.11

STEP 10 Insert four connector pegs and two cross connectors as shown in Figure 6.12.

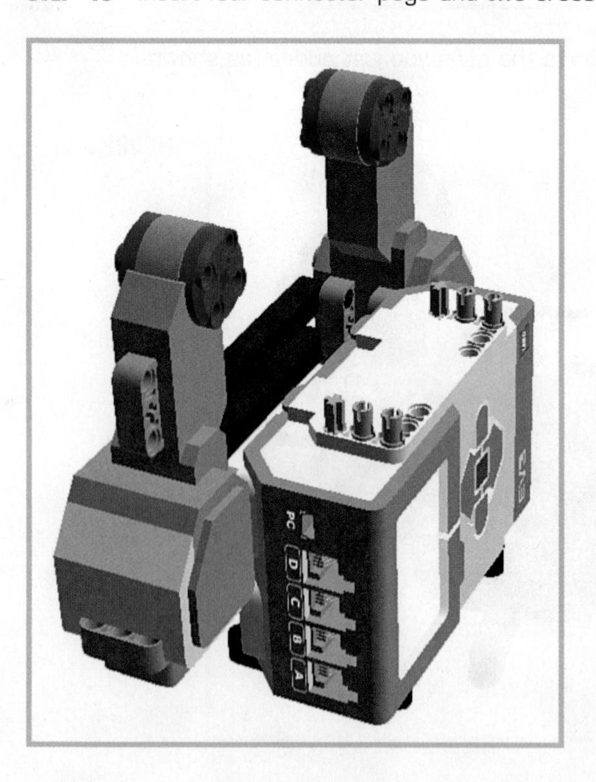

FIGURE 6.12

STEP 11 Attach two 2M and two 3M cross blocks. It should look like Figure 6.13.

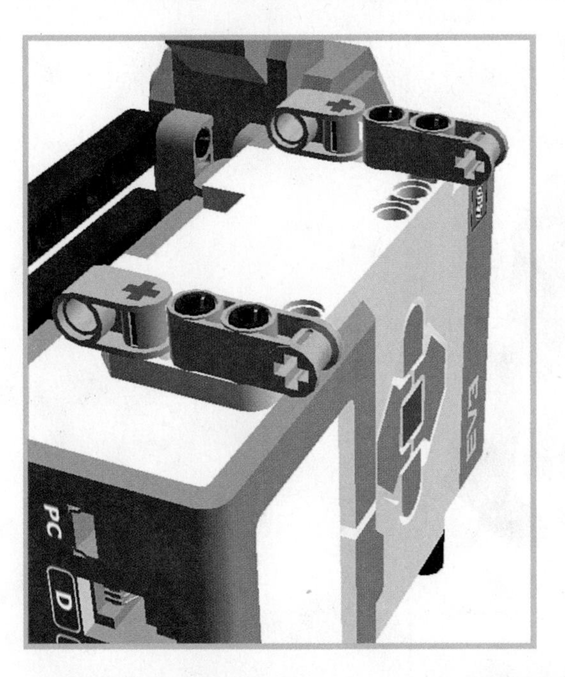

FIGURE 6.13

STEP 12 Let's switch things up a bit and work on the main chassis of the robot. Grab a pair of 5x11 chassis bricks and add a bunch of connector pegs as shown in Figure 6.14.

FIGURE 6.14

STEP 13 Connect the two 5x11 chassis bricks with a 5x7 chassis brick (see Figure 6.15).

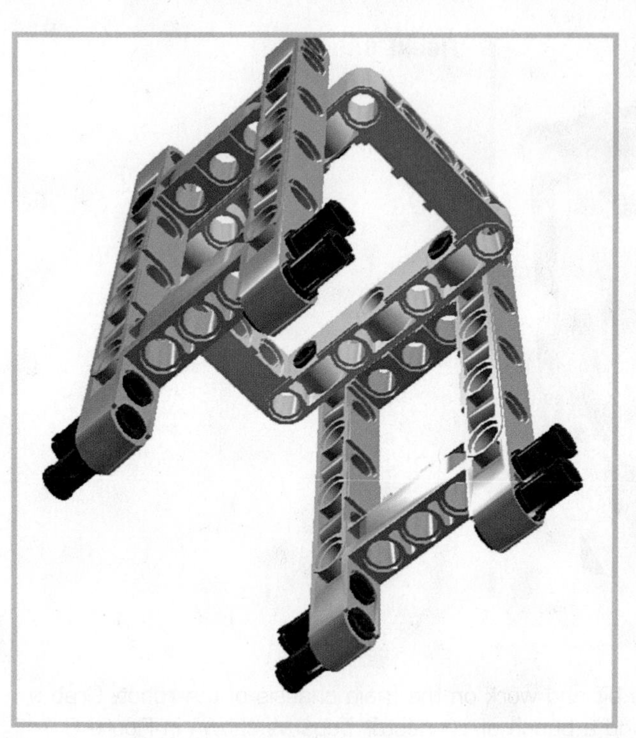

FIGURE 6.15

STEP 14 Add four T-beams to the chassis bricks as shown in Figure 6.16.

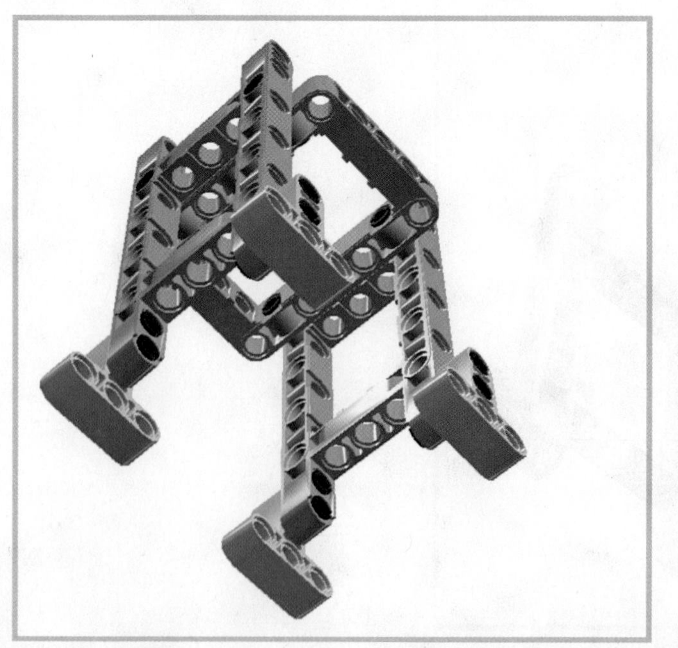

FIGURE 6.16

STEP 15 Insert eight connector pegs to the bottom of the T-beams, as shown in Figure 6.17.

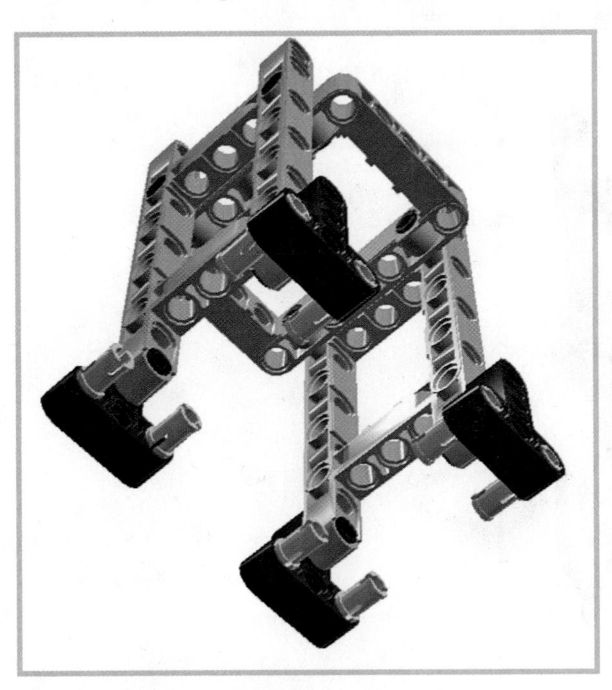

FIGURE 6.17

STEP 16 Add a pair of 9M beams to help stabilize the assembly (see Figure 6.18).

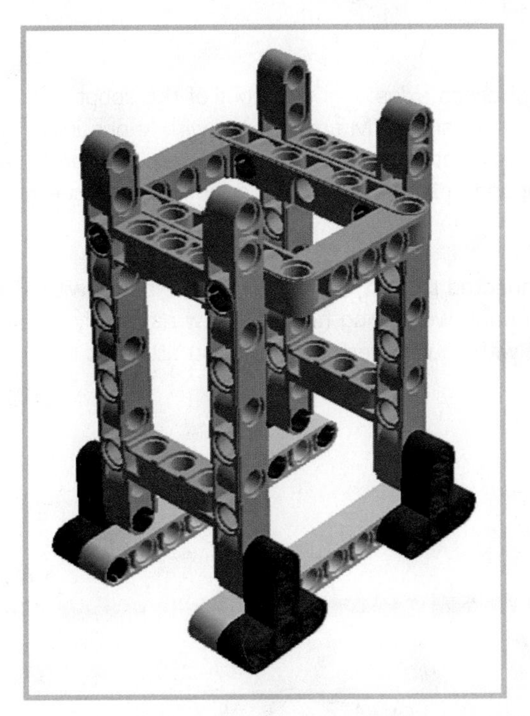

FIGURE 6.18

STEP 17 Set the assembly you just built on top of the Intelligent Brick, as shown in Figure 6.19. Don't worry, you secure it the next step!

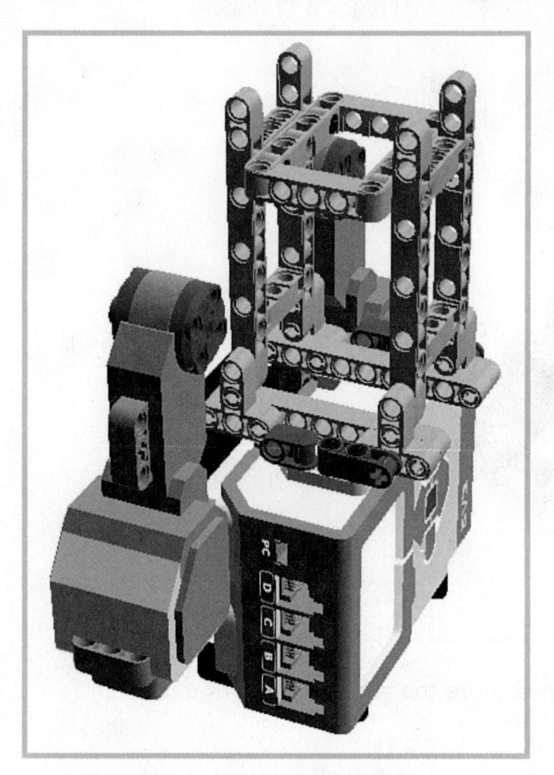

FIGURE 6.19

STEP 18 Secure the chassis with a series of cross axles. For the front of the robot (where you find the screen on the EV3), insert a pair of 3M cross axles with knobs into the cross blocks. The back of the chassis is secured with two 4M cross axles with end stops, with a 180-degree angle element trapped in between. It should look like you see in Figure 6.20.

STEP 19 Next, add a pair of 7M beams connected to each other with a 3M beam with snaps. As you look at Figure 6.21 you're probably wondering how this new assembly attaches to the main one. Just hold it there with your fingers for now; you attach it in the next step.

FIGURE 6.20

FIGURE 6.21

STEP 20 Secure the 7M beams with a pair of 8M cross axles with end stops. Note that one axle goes in one way, and the other axle inserts from the opposite side. Also note the pair of half bushings serving as spacers toward the front of the robot. Figure 6.22 shows how it should look.

FIGURE 6.22

STEP 21 Slide bushings onto the ends of the 8M cross axles you just added, as shown in Figure 6.23.

STEP 22 Let's insert eight connector pegs. Put two in each motor's hub, as shown in Figure 6.24, plus two in the horizontal beams of both of the 3x11 chassis bricks.

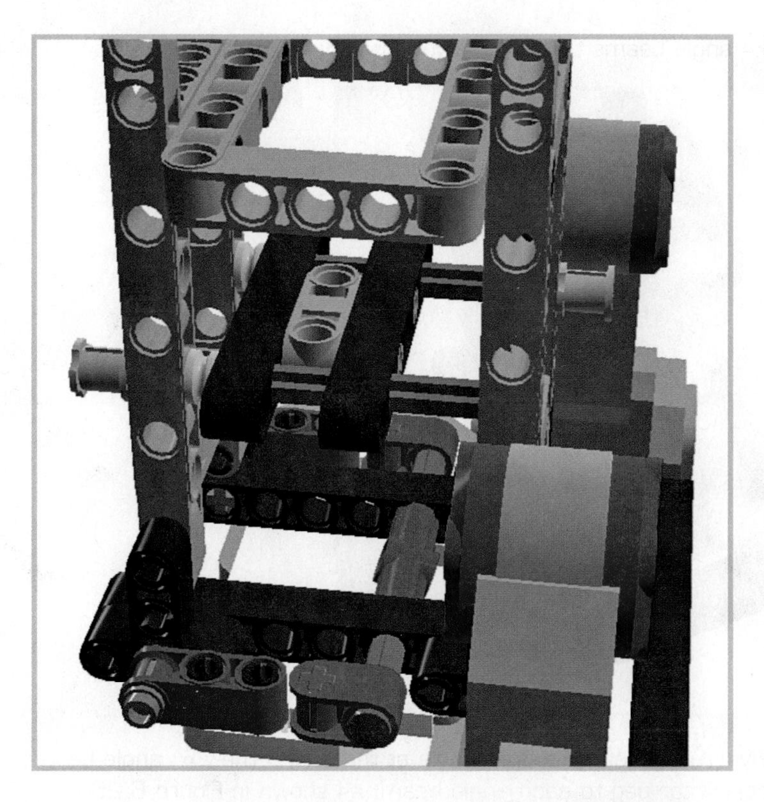

FIGURE 6.23

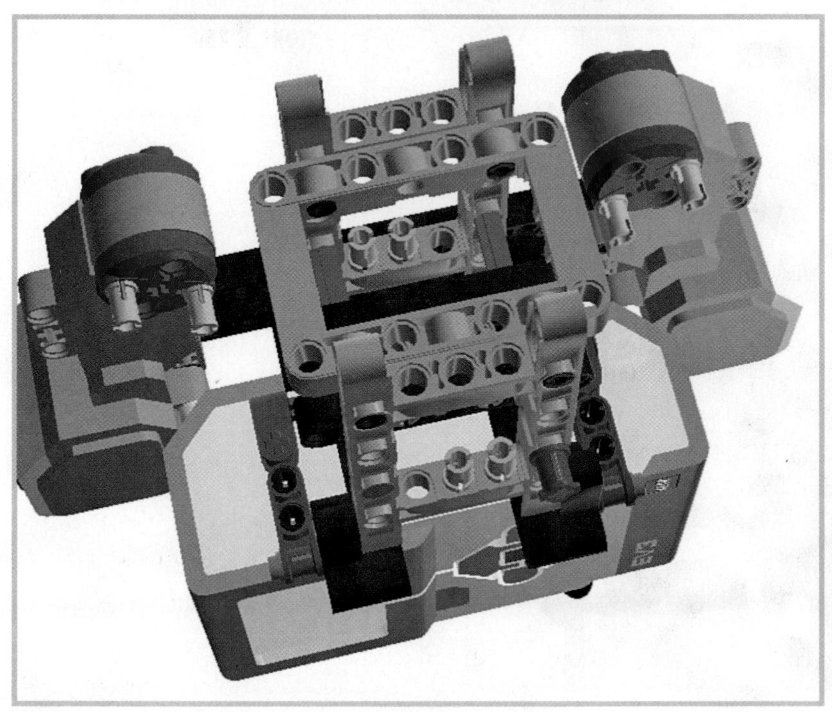

FIGURE 6.24

STEP 23 Attach two 2x4 angle beams to the pegs you just added (see Figure 6.25).

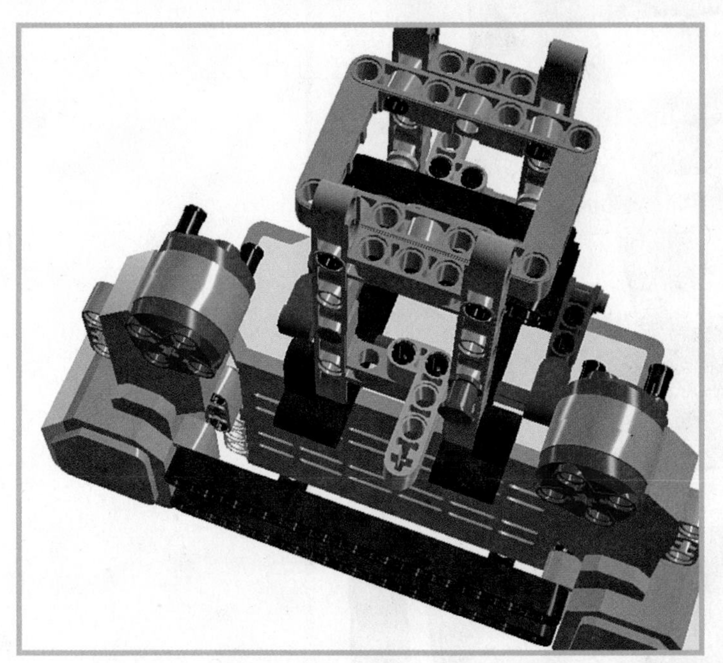

FIGURE 6.25

STEP 24 Insert a red 2M cross axle in the cross hole at the end of the 2x4 angle beams. Then add a blue cross connector peg to each angle beam, as shown in Figure 6.26.

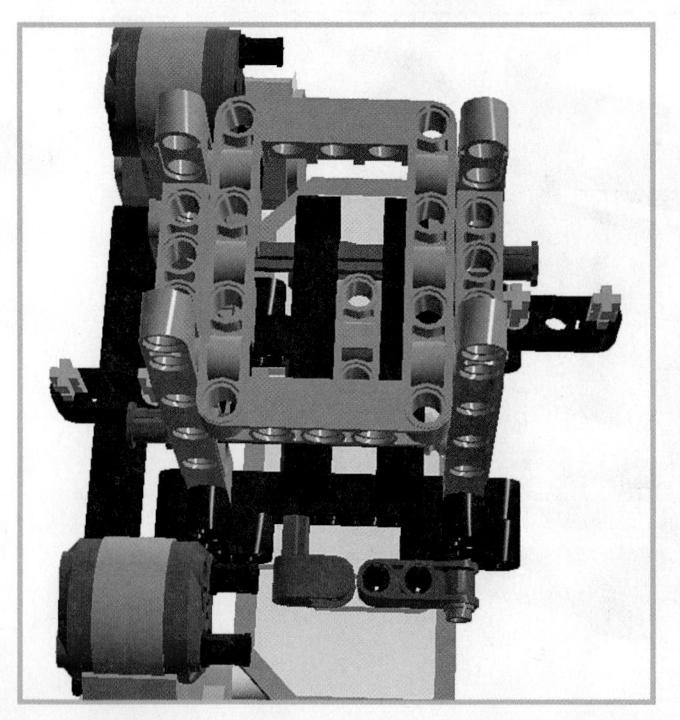

FIGURE 6.26

STEP 25 Add six cross-connectors: Two go in each 5x11 chassis brick. The pegs visible on the back of the assembly mirror those in the front (see Figure 6.27).

FIGURE 6.27

STEP 26 Add double cross blocks to the cross connector pairs you just placed on the chassis bricks. You can see how it should look in Figure 6.28.

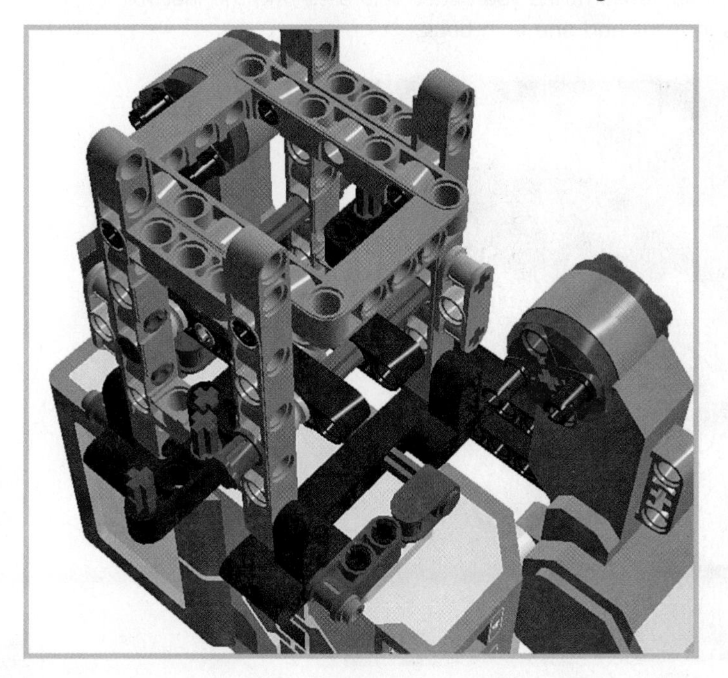

FIGURE 6.28

STEP 27 Slide an 8M axle with end stop through one of the double cross blocks you just added, trapping a Z20 gear and a 3x7 angle beam along the way, as shown in Figure 6.29. Secure the end with a bushing.

FIGURE 6.29

STEP 28 Rotate the model and repeat the process on the other side, as shown in Figure 6.30. Note that the gear assemblies you added this step and the last mirror each other, so one gear is in front and one is in back.

FIGURE 6.30

STEP 29 Next, we work on a new assembly. Slide a double cross block onto a 6M axle, as shown in Figure 6.31. Secure the double cross block with a bushing on one side and a half bushing on the other.

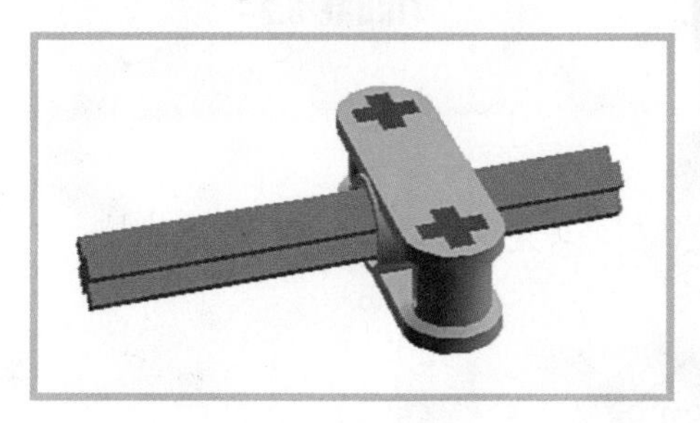

FIGURE 6.31

STEP 30 Slide a 3x7 angle beam onto the axle, as shown in Figure 6.32.

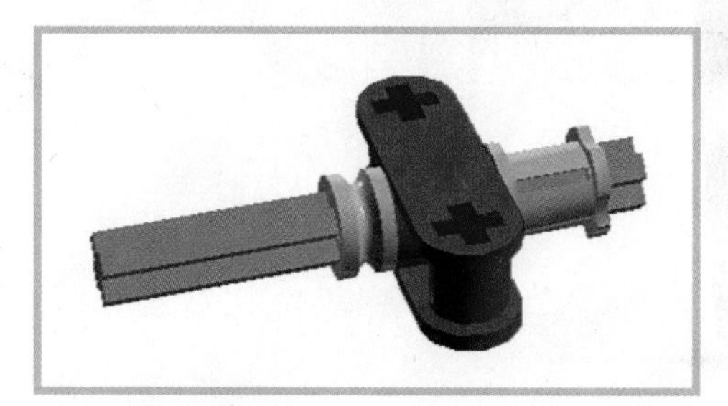

FIGURE 6.32

STEP 31 Finish off the assembly with a Z12 gear on the end (see Figure 6.33). Now, make a second assembly just like the first one! You want two of these assemblies.

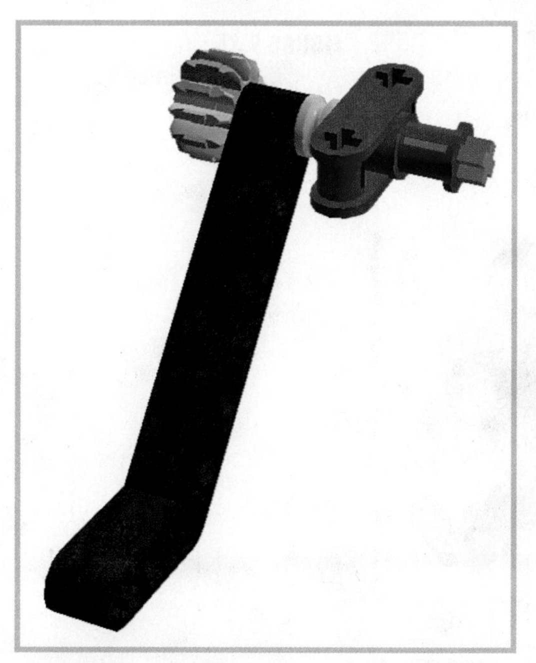

FIGURE 6.33

STEP 32 Attach the two gear assemblies you just built to the 2x4 angle beams sticking out of the chassis in front and back. Figure 6.34 shows how it should look.

FIGURE 6.34

STEP 33 Next, snap a pair of 7M beams onto the motor hubs, as shown in Figure 6.35. While you're in the neighborhood, insert a pair of gray no-friction pegs to the ends of the beams.

FIGURE 6.35

STEP 34 Attach 5M beams to the gray pegs on the ends of the 7M beams you just added, allowing them to swing loosely (see Figure 6.36).

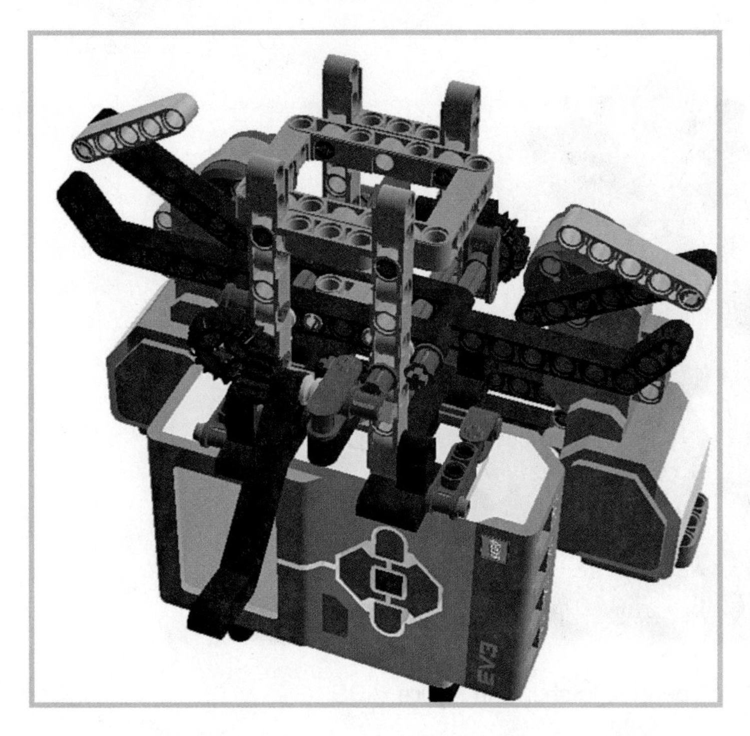

FIGURE 6.36

STEP 35 Insert a dozen connector pegs into the chassis bricks, as shown in Figure 6.37.

FIGURE 6.37

STEP 36 Attach four cross block forks to the chassis bricks (see Figure 6.38). The petal assemblies you created will attach to the forks in a future step.

FIGURE 6.38

STEP 37 Insert a pair of 5M beams as shown in Figure 6.39.

FIGURE 6.39

STEP 38 Let's add the two smaller petals (the ones built with the 3x7 panels) to the front and back forks. The red 11M beam attaches to the fork, secured with a 3M axle, while its support beam connects to the 3x7 angle beam. Figure 6.40 shows how it should look.

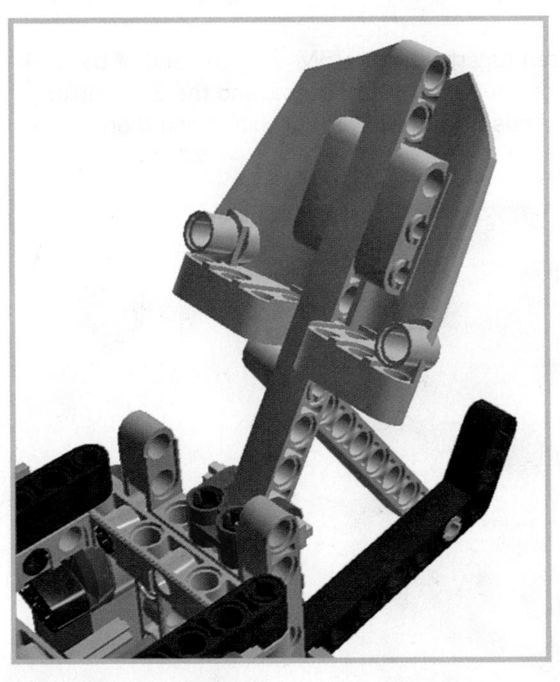

FIGURE 6.40

STEP 39 Attach the left and right petals, which have the longer 3x11 panels. The red beams attach to the remaining two forks using 3M axles, as shown in Figure 6.41.

FIGURE 6.41

STEP 40 Finally, we need to secure the support beams for the big petals. This is kind of a complicated step and I had a hard time illustrating it in my design program, so instead I'm showing a photo of the robot that demonstrates how to connect things (see Figure 6.42).

There are three beams that get connected together with a 5M axle: the end of the 5M beam you added in Step 35, the end of the petal's support beam, and the 3x7 angle beam you added in Step 31. Secure the ends of the axle with bushings and then repeat the process with the final petal. You're done!

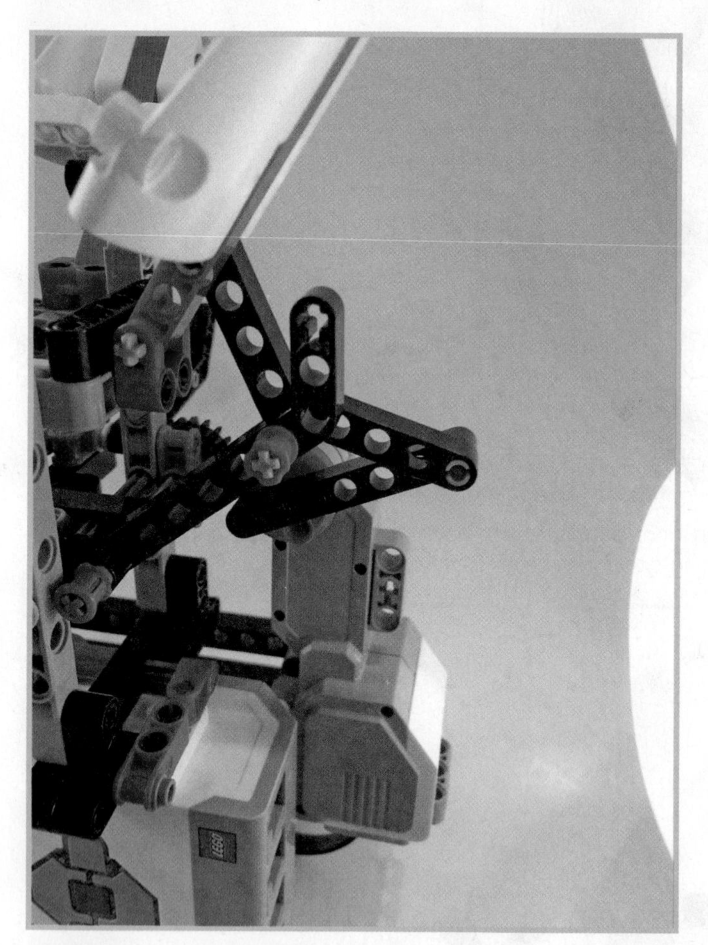

FIGURE 6.42

Program the Robot Flower

Once the build is complete, you can focus on writing the Mindstorms program.

STEP 1 Drag a Variable block out of the palette and place it as you see in Figure 6.44. Name the variable "hours," give it a default value of 0, and make the variable Write and Numeric.

FIGURE 6.43

STEP 2 Drop in a Loop block as shown in Figure 6.44. Keep the default loop name of 01.

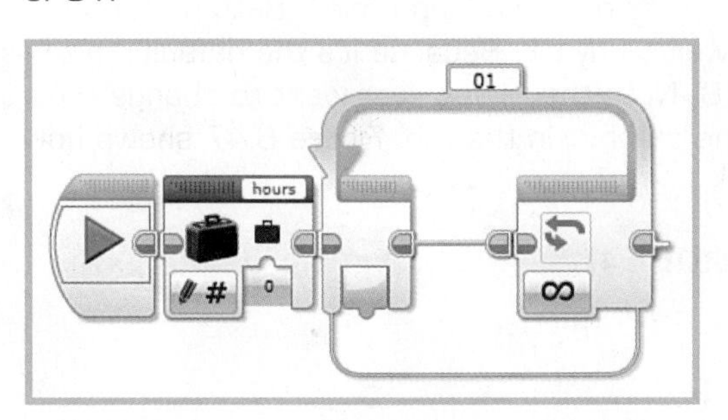

FIGURE 6.44

STEP 3 Place three blocks in your newly created loop: First, add another Variable block, though set this one to Read. Next, add a Logic block set for addition with the first number pulled from the Variable block and the second number set at the default of 1. Finally, drop another Variable block set to Write, with the input coming from the output of the Logic block. These three blocks increment the variable hours by adding 1 to it every time the loop cycles. Figure 6.45 shows how it should look.

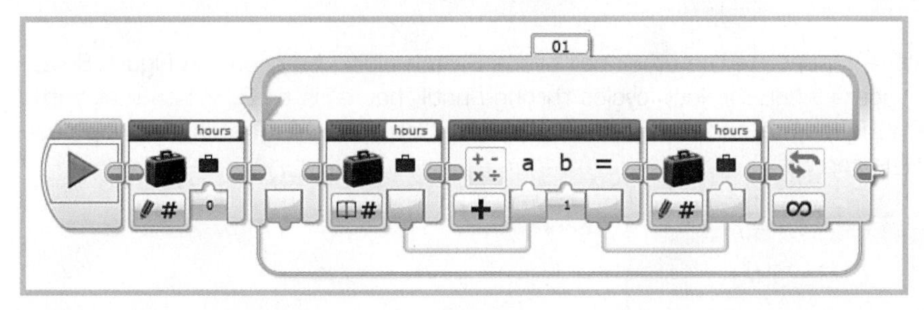

FIGURE 6.45

STEP 4 Next, drop a Display block into the loop as shown in Figure 6.46. This is purely for diagnostic purposes and merely displays the current value of "hours" on the EV3's screen. You can totally ditch this block without affecting anything. Set the Display block to Text-Pixels and then click on the default text value ("Mindstorms") to Wired, allowing you to display the result of the Logic block's computations with the help of a data wire.

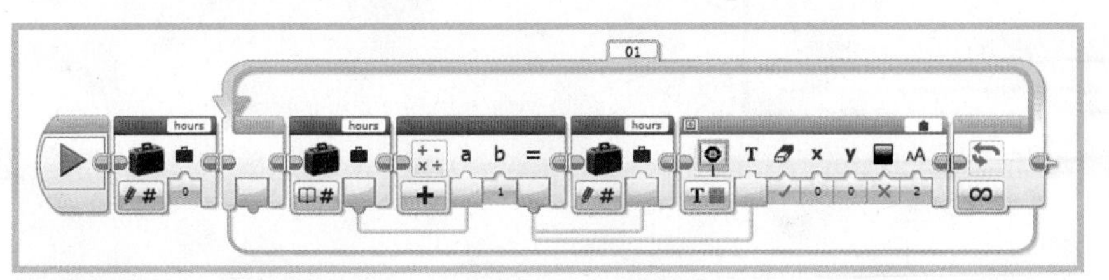

FIGURE 6.46

STEP 5 Drag a Switch block from the palette and drag it into the loop. You want tabbed viewing, Numeric, with one cell set to 1 (with default selected) and one cell set to 8. Drag in another wire from the Logic block to tell the Switch the current "hours" value.

In the "1" cell, add a Move Tank block and set it to 25% power on both, with rotations set to -0.25. The Move Tank block controls two motors simultaneously, which works great for us.

This is the default action that takes place every time the loop cycles: Both motors turn a quarter revolution, opening the flower a tiny bit. Because it's the default, this cell triggers anytime the variable *isn't* an 8. Note that if you ever want to change the number of loops (hours), simply change the number in the tab. Figure 6.47 shows how it should look.

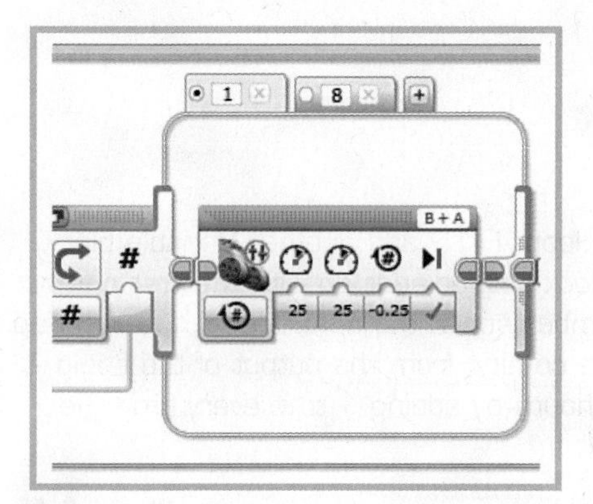

FIGURE 6.47

STEP 6 Next, let's populate the other cell of the Switch block, as shown in Figure 6.48. The "8" cell triggers when the loop cycles through until "hours" is an 8. When that happens, we want the loop to end. Drag a Loop Interrupt into the cell. The default loop designation of 01 is correct.

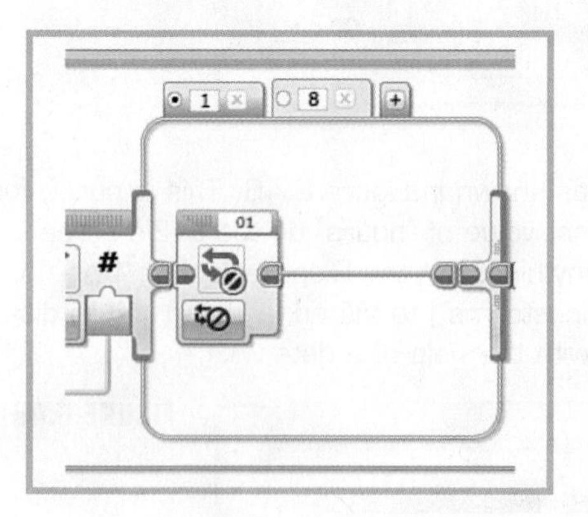

FIGURE 6.48

STEP 7 The loop is not yet complete! Let's add one more block: a Timer, shown in Figure 6.49. The Timer delays the robot's movements so that they take place once an hour.

Set it to Change - Time, and then put 5 down for the duration. But wait, I said an hour! Ever try debugging a robot when it only moves once an hour? The default of 5 seconds suffices for now, and when you have the flower working the way you want, change the duration to 3600, the number of seconds in an hour. You're done with the loop!

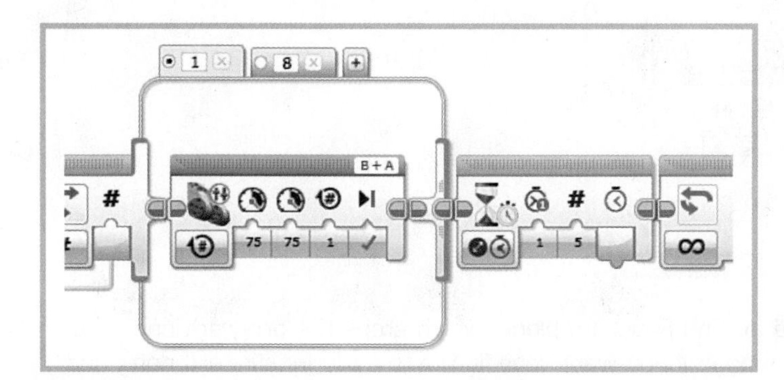

FIGURE 6.49

STEP 8 The loop you just completed opens the flower throughout the course of 8 hours. The next step is to make it close again. Duplicate the loop by selecting it (the boundaries turn blue) and cutting and pasting it next to the other, as you see in Figure 6.50. Change its name to 02.

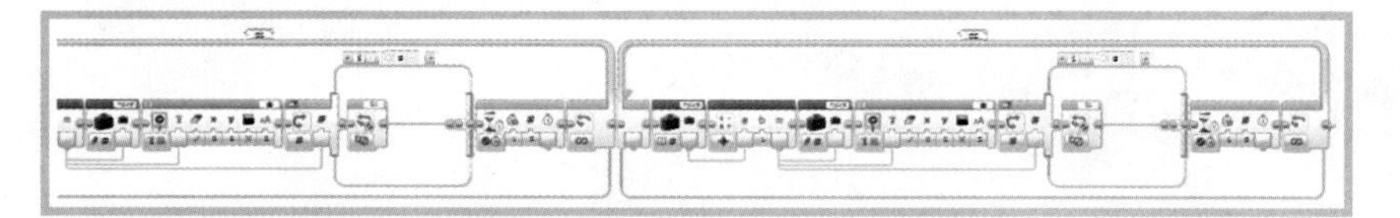

FIGURE 6.50

STEP 9 Obviously we need to change the settings so the motors move in the opposite direction. First, however, let's change the Logic block for the new loop so that it decrements—subtracts 1 with each loop cycle. Simply change the setting to Subtract. The plus turns into a minus, as shown in Figure 6.51.

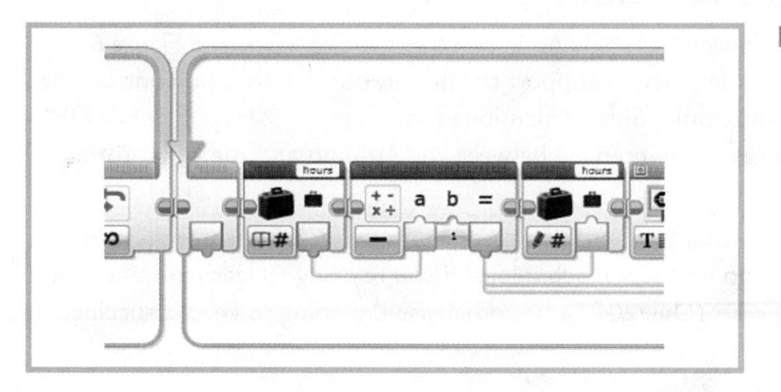

FIGURE 6.51

STEP 10 Now for the Switch block, shown in Figure 6.52. First, renumber the two tabs, so 1 becomes 8 and vice versa. Since the variable hours is counting down, 8 becomes the default and 1 is the end. The next thing you need to do is change the motor's rotation by giving it a positive rotation number. Finally, the Loop Interrupt should be changed to interrupt Loop 02.

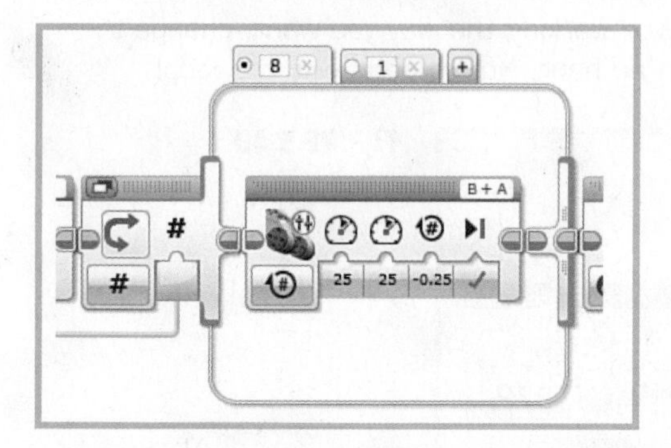

FIGURE 6.52

STEP 11 Finally, let's place an End Program block, which stops the program once the second loop has run. It's optional; if you want your flower to cycle indefinitely, don't add it. Put the End Program block outside the second loop, as shown in Figure 6.53.

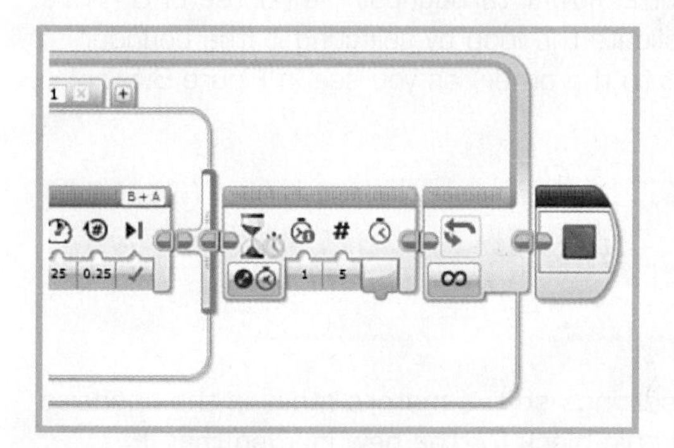

FIGURE 6.53

Substituting the Arduino

Let's mix things up big time by using an Arduino in place of the EV3 brick (see Figure 6.54). Oh, we leave the physical brick in place to support the flower, but the Arduino controls the motors with the help of a Bricktronics Shield, mentioned in Chapter 5, "Hacking LEGO II: Alternate Controllers." It serves as an interface between the Arduino and the Mindstorms motors.

However, I'm going to begin with a brief Arduino tutorial. This isn't a book about Arduinos, however, so I'm not going to go into too much depth. If you're dying to learn more, check out my book *Arduino for Beginners* (Que 2013), but otherwise I'm going to keep it succinct.

FIGURE 6.54 You learn how to control the Robot Flower with an Arduino.

Quick and Dirty Arduino

Here's what you need to get going:

Read this How-To

Two solid documents to help you learn about the basics:

http://www.arduino.cc/en/Main/Howto

http://www.arduino.cc/en/Tutorial/Foundations

Buy an Arduino

You can purchase an Arduino UNO at SparkFun.com, P/N 11021 among many other sites and physical stores.

Learn About the Board

Learn more about the UNO and its components:

http://www.arduino.cc/en/Main/ArduinoBoardUno

Download the Environment

You need to download the platform's development environment, available for Mac, Windows, and Linux:

http://www.arduino.cc/en/Main/Software

Learn About the Environment

Once you have it installed, familiarize yourself with its features:

http://www.arduino.cc/en/Guide/Environment

Parts

To equip your Robot Flower with an Arduino, you need the following parts:

- **Arduino UNO**—As mentioned, they are widely available.
- **Bricktronics Shield**—Shown in Figure 6.55, available at http://www.wayneandlayne.com/bricktronics/.
- **Power supply**—SparkFun has a good one, P/N 298.

FIGURE 6.55 The Bricktronics Shield.

Steps

Let's go through the quick and simple setup of the Arduino project:

STEP 1 Plug the Bricktronics Shield on top of the Arduino, inserting the shield's header pins into the Arduino's female headers. It should nest neatly, as shown in Figure 6.56.

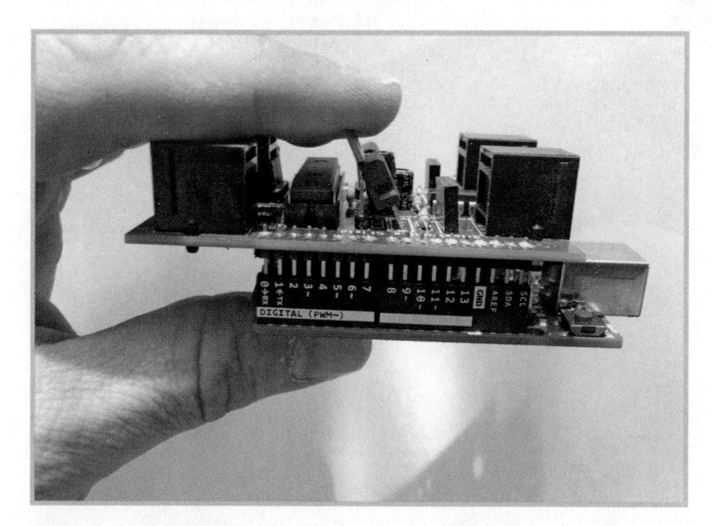

FIGURE 6.56

STEP 2 Plug the Mindstorms motors into the Bricktronics motor ports. These are the two ports by themselves, shown in Figure 6.57.

FIGURE 6.57

Programming the Arduino

Upload the following sketch to the Arduino. When the code has loaded, the board automatically launches the program.

```
#include <Wire.h>
#include <Bricktronics.h>

Bricktronics brick = Bricktronics();
PIDMotor l = PIDMotor(&brick, 1);
PIDMotor r = PIDMotor(&brick, 2);

void setup() // arduino sketches have a Setup, which runs once,
//and a Loop, which runs until the loop is broken.
{
  Serial.begin(115200);
  brick.begin();
  l.begin();
  r.begin();

}

void loop()
{
  switch(hours) // switch works much like EV3's Switch block.
  case 1:  //the flower opens

  l.set_speed(-75);
  r.set_speed(-75);
  delay(150);
  l.set_speed(0);
  r.set_speed(0);
  delay(1000);

  ++hours; // increments the hours

  if (hours=9) {
    case = 2;
  }

  break;

case 2:  //the flower closes
```

```
  l.set_speed(150); //note the different settings vs. opening the flower.
//the robot needs more power to open vs. close, because it's fighting gravity.
  r.set_speed(150);
  delay(150);
  l.set_speed(0);
  r.set_speed(0);
  delay(360000); // pauses the program for 1 hour

  --hours; // decrements the hours

  if (hours=0) {
    case = 1;
  }

  break;

}
```

Summary

Whew! In this rather long chapter, you had the opportunity to make a time-dependent robot flower model and program it with both Arduino and Mindstorms controllers. In Chapter 7, "Hacking LEGO III: Create Your Own LEGO Parts," you explore one of my favorite aspects of LEGO hacking: creating your own beams, plates, and gears to interface with Mindstorms' already robust assortment of elements.

Hacking LEGO III: Create Your Own LEGO Parts

A friend of mine was printing up LEGO bricks at a convention when a wide-eyed kid asked him if he could print bricks that the LEGO Group doesn't provide in its sets. Such as, the kid speculated, a 5x3 brick? The answer was, absolutely! Someone already has: Jorg Janssen posted a 5x3 brick on Thingiverse.com that can be customized and output with a 3D printer (see Figure 7.1). You can find it at http://www.thingiverse.com/thing:178627.

The LEGO Group makes an effort to provide a robust assortment of elements, both commonplace and rare, to give you the best possible robot-building experience. LEGO's Mindstorms sets have a bunch of parts, and its line of Technic motorized models provides still more. In the end, however, LEGO can't make every part imaginable. Sooner or later you'll need something that doesn't exist. Maybe you need a part with a hole configuration different from other Mindstorms parts, or one that has alternate angles, or something bigger or smaller than typical parts. When that happens, you can certainly redesign your model—in fact, an influential plurality of LEGO fans advocate doing exactly that: working solely within the confines of what LEGO provides. If we all felt that way, however, this book wouldn't exist!

This chapter details some experiments with creating new parts, mostly using computer-controlled, precision tools: a laser cutter and 3D printer. While out of the reach of many people, these tools are a necessity in creating elements precise enough to work seamlessly with existing parts. They work by following a path described in a long text document: Move the toolhead this direction and this far and then move it in another direction. Computer utilities are used to turn computer graphics into these instructions.

FIGURE 7.1 A 5x3 brick doesn't exist in any official LEGO set. So print one! Credit: Jorg Janssen (Creative Commons).

Designing Your Own Parts

If you can't download someone else's design, you need to create and output your own—but wouldn't that be more fun anyway? As I mentioned, you typically need computer-controlled tools to make elements precise enough to be usable. You can see a design file in Figure 7.2. The file outputs several Mindstorms-compatible wooden plates, described later in this chapter.

FIGURE 7.2 The laser cutter follows the paths shown on the screen.

Check Your Dimensions

All LEGO parts follow a certain pattern, allowing you to be reasonably sure that two dissimilar parts from different sets will snap together with ease. This, combined with computer-controlled tools, allows you to create your own parts—whether out of plastic, wood, or metal.

Figure 7.3 shows the hole pattern of a LEGO Technic beam, such as are found in the Mindstorms EV3 set. However, to summarize:

Holes are 4.2mm in diameter and are set into small indentations 0.8mm thick, making the overall width 5.8mm. Beam walls are 0.8mm thick, making for an overall beam width of around 7.5mm. Beam thickness is 7.8mm and half-width parts (such as the yellow half bushings) are 3.9mm thick.

If you're a math nerd and want to jump feet first into all the glorious ways LEGO dimensions are awesome—and there are many; it's a carefully designed system—check out Robert Cailliau's "Lego Dimensions" page, packed with notes on every possible permutation of LEGO elements' measurements (http://www.robertcailliau.eu/Alphabetical/L/Lego/Dimensions/General%20Considerations/).

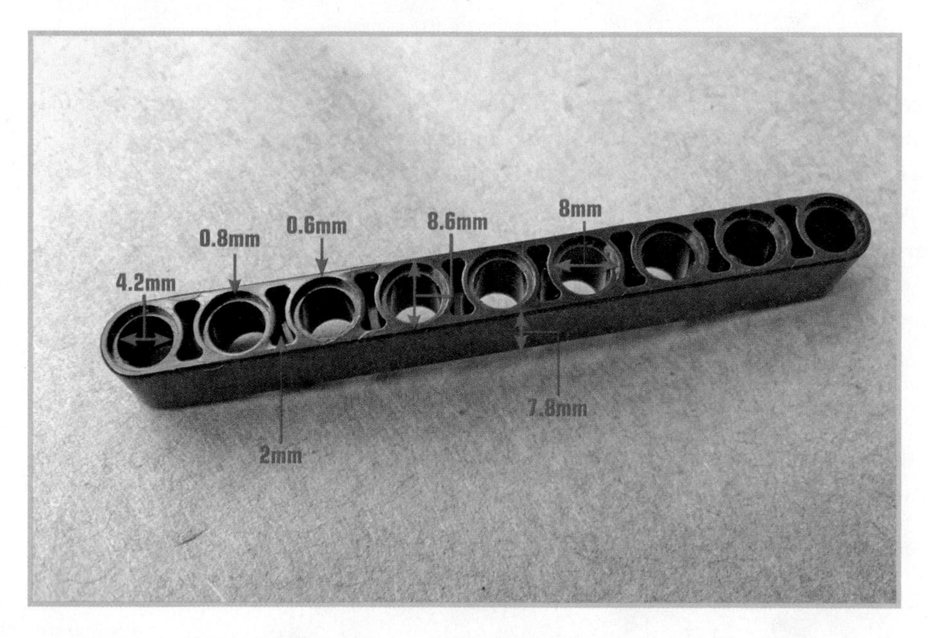

FIGURE 7.3 LEGO beams' hole spacing follows a specific pattern.

Create the Design File

Once you have your dimensions figured out, launch some software and build the part. Easier written than done! What software do you choose? Here are some packages I employ to work with vector and 3D art. There are many others besides, but these are the two I use on a regular basis.

Inkscape

This vector art program compares well to Adobe Illustrator, the industry standard in vector design. Inkscape is free and open source, meaning it doesn't cost you any money to use, and if you had bug fixes or improvements, and were a programmer, you could actually participate in making the program better.

Inkscape works by drawing a series of curves and lines, anchored together to form shapes. You can create polygons and circles at will, or create and clone more complicated shapes. Illustrator may be more polished and have more powerful features, but it's also $800!

I use Inkscape for doing the initial designs for laser cutting and CNC (computer numeric control) milling. You can download Inkscape at https://inkscape.org/.

SketchUp

I use Inkscape for 2D work and SketchUp for 3D work. SketchUp works in vectors the same way but with the added ability to work in three dimensions. If you were to draw a square, for instance, you could grab the edge and "pull it up" to form a cube.

SketchUp, shown in Figure 7.4, is also free—at least the simplest version is free. However, the Pro version, which sells for $590, has some cool features, such as superior importing of third-party files, better measurements, better computing of area and volume, and a widget that converts your design into presentation slides.

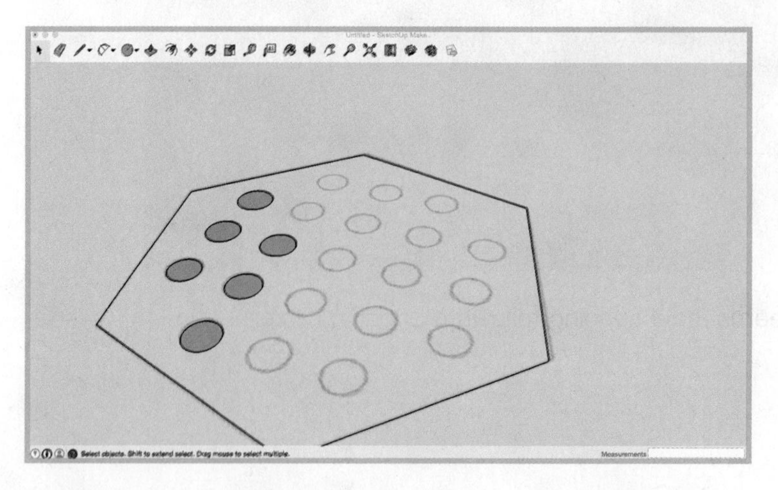

FIGURE 7.4 Building a Mindstorms part in SketchUp.

Output and Iterate

The only way to be *really* sure that you have the right part is to make one and add it to your robot. This means, inevitably, that you end up with a number of experiments, mistakes, and rejects that inspire you to build a better part (see Figure 7.5)!

Don't be frustrated if (when!) your first attempt doesn't work out. Part of the process of creating something new is trying different techniques and configurations until you find the right one. This is called *iteration*, and it can actually be a lot of fun!

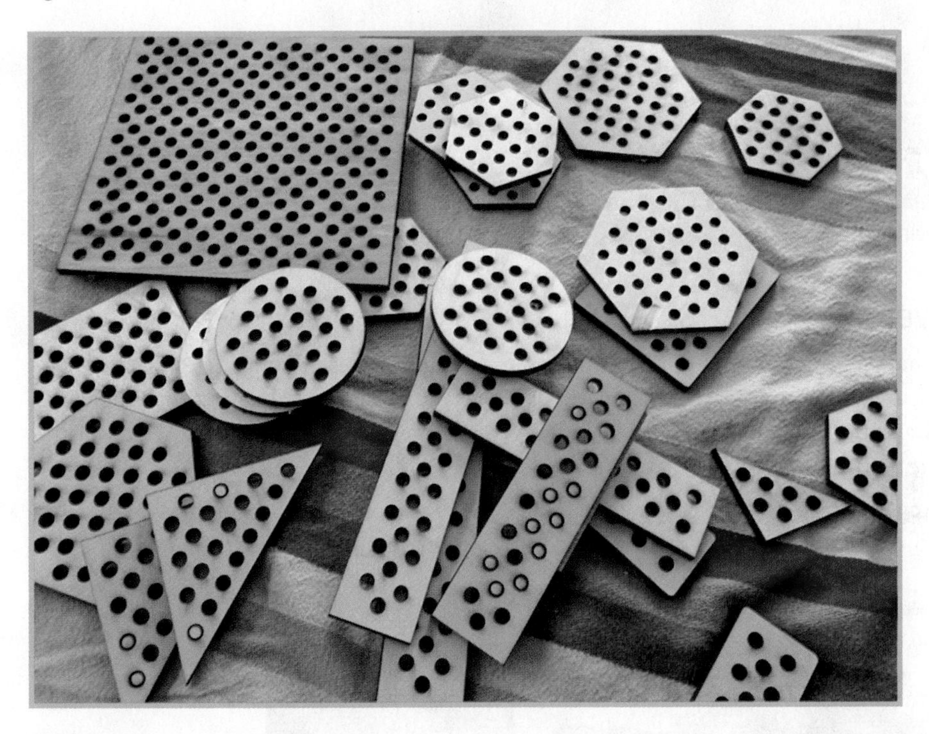

FIGURE 7.5 Once you have output your design, test it to see whether it works.

Finding Designs

If you need a specific element, chances are someone else has needed it as well. There are a number of outstanding sites where LEGO hackers share their projects, allowing visitors to download their designs. I always check online before I take the time to build a new part. Some designs are obscure, but others are no-brainers. For instance, Steve Medwin's curved beam design (shown in Figure 7.6) allows you to create Technic beams with curves, making up for the fact that LEGO only sells straight and angled beams. (You can download his design file from Thingiverse: http://www.thingiverse.com/thing:629875.)

Let's go over a number of sites where existing designs can be downloaded.

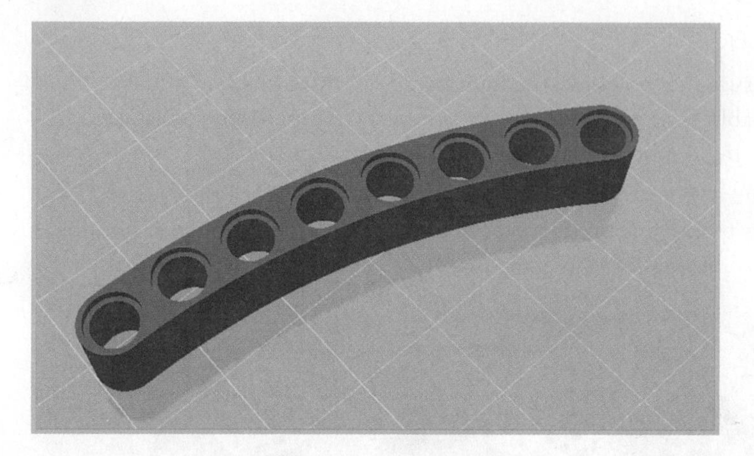

FIGURE 7.6 Ever need a curved beam in a design? Credit: Steve Medwin, 3dmedwin.com.

Thingiverse

My personal go-to site for 3D printing and laser-cutting resources, Thingiverse is run by Makerbot Industries, a company selling 3D printers, and as such it mostly offers designs for 3D printing, but if you're looking for laser designs, Thingiverse has some of those as well (see Figure 7.7).

Each "thing" (design) is detailed on its own page, with visitors able to comment on it, download the design, or post "makes" that show what you did with the files.

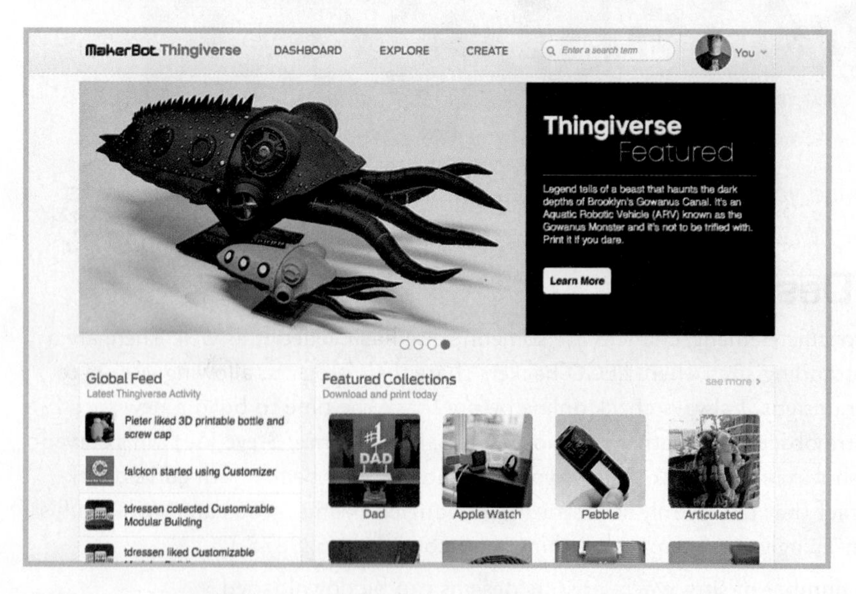

FIGURE 7.7 Thingiverse is a popular destination for 3D printing fans.

LDraw

LDraw is the effort by LEGO fans to catalog every single LEGO element that has been released, providing 3D scans of each part so that anyone can print any LEGO part (see Figure 7.8). It's a bold plan, made even bolder by the creation of parts that "don't exist but should." Unlike Thingiverse, there is a community of editors who ensure that only the best designs are posted, so you can be confident every part is both unique and well-designed.

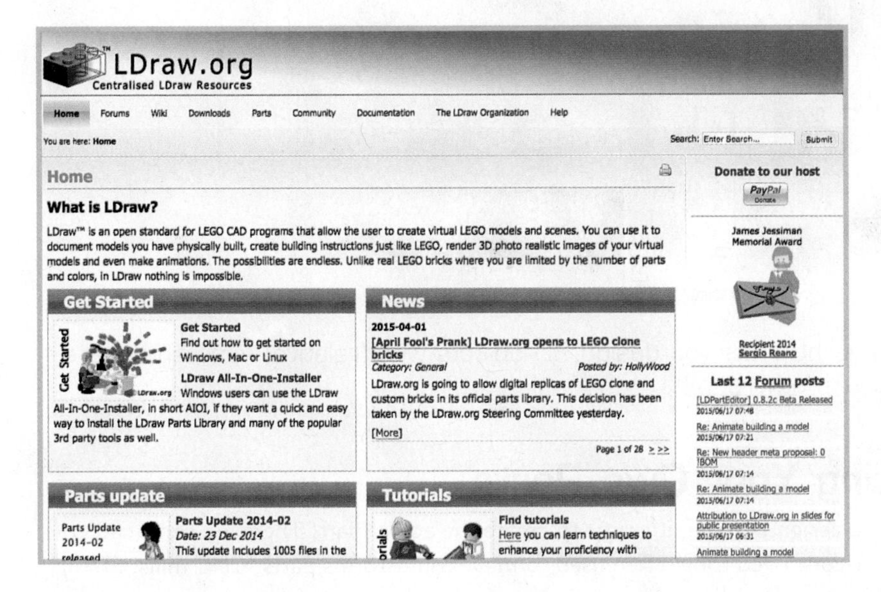

FIGURE 7.8 LDraw contributors carefully document every official brick and a lot of unofficial ones!

SketchUp Warehouse

I already mentioned SketchUp, a dirt-simple 3D rendering program. SketchUp wisely includes a library of parts you can download and add to your creation. So for instance, if you're making an architectural drawing of a building, you could download car shapes to decorate the street next to the structure.

SketchUp's library is called the Warehouse, and you can find many examples of LEGO-related designs to download: https://3dwarehouse.sketchup.com/search.html?q=mindstorms (see Figure 7.9). Although the SketchUp file type is not natively compatible with most 3D printers, you can easily export your files to the .STL format, which is.

FIGURE 7.9 SketchUp lets you design 3D objects with (relative) ease and share your designs with others.

Outputting Your Own Parts

Once you have a design in hand, it's time to create the actual part! There are three categories of computer-controlled tools often used to make Mindstorms parts: CNC mills, 3D printers, and laser cutters.

CNC Mill

CNC stands for computer-numerically controlled, meaning the tool follows a series of instructions from a computer to move its toolhead. A CNC mill consists of a rotary tool like a Dremel that, with the help of a variety of bits, grinds out material to form a shape from a block of wood, metal, or plastic. This technique is called "subtractive" because it takes material out of a block to form a shape, much like a stone sculptor chisels away at a marble block to make a statue.

You often hear "CNC mills" shortened to "CNC," though technically 3D printers and laser cutters use similar control methodologies and could all be described as computer-numerically controlled. That said, the term "CNC" by itself typically refers to mills and similar devices, like the computer-controlled router pictured in Figure 7.10.

The following are a couple of LEGO-related CNC projects I've found online.

FIGURE 7.10 A CNC router stands ready to cut.

Klann LEGO Spider

The Klann linkage simulates the gait of a crab or spider, using mechanical linkages that allow you to raise one leg while another is lowering (see Figure 7.11). Not only is this a cool way to make a robot move around (compared to, say, wheels), but it also looks very animal-like.

The robot's builder, Mark Weller, milled the legs out of 3/8-inch plastic, but the rest of the components are Mindstorms-standard axles and bushings. You can download the design at: http://www.thingiverse.com/thing:1643.

FIGURE 7.11 This Mindstorms robot uses Klann linkage legs.
Credit: Mark Weller.

Aluminum LEGO

Dutch tinkerer Tristram Budel output the gorgeous aluminum bricks pictured in Figure 7.12, milled out of blocks of solid metal. Though Tristram's aluminum LEGO bricks are "System" elements, meaning traditional studded LEGO rather than Mindstorms, you can see how the same techniques could be used to make aluminum Mindstorms parts. You can learn how to make them on Instructables: http://www.instructables.com/id/DIY-LEGO/.

FIGURE 7.12 The classic 2x4 LEGO brick, carved out of metal. Credit: Tristram Budel.

Laser Cutter

While known primarily for 2D outputs—the laser cuts shapes out of wood and plastic—laser cutters can still play a role in LEGO building. They create great interface pieces that can join two dissimilar components, such as helping an electronic sensor attach onto a Mindstorms robot.

Laser cutters, such as the one in Figure 7.13, follow the same general parameters as other CNC tools but use a cutting laser to do their work. Most consumer lasers can't cut through metal or certain other materials but can easily carve plywood, acrylic sheets, and so on.

Let's go over a couple of cool laser-cutter projects.

FIGURE 7.13 The toolhead of the laser cutter follows its path.

LazerStorm

A notable omission in the Mindstorms kit are big mounting plates onto which other components can be attached. LazerStorm (https://www.tindie.com/products/n1/lazerstorm-elite-pack/) is a project consisting of several different shapes, lasered out of plywood. You can see some examples in Figure 7.14. There are circles, rectangles, hexagons, and so on, each peppered with Technic-spacing mounting holes.

The plates can be used for bases, or simply for structural support for making our constructs sturdier. Because they're made of wood, you can drill into the plates and further modify them—such as by attaching a sensor or microcontroller.

FIGURE 7.14 LazerStorm plates feature Technic-spacing mounting holes.

Arduino to LEGO Interface Pieces

This was an early attempt (2009!) at creating mounting hardware for LEGO that allows non-LEGO components to mate with it. In this case, Stuart McFarlan designed microcontroller and servo mounting plates that allow you to use a LEGO chassis with non-LEGO motors and other components (see Figure 7.15). You can learn more about Stuart's projects at oomlout.com or download the files from http://www.thingiverse.com/thing:403.

FIGURE 7.15 LEGO mates with non-LEGO with the help of laser-cut acrylic. Credit: Stuart McFarlan.

BrickPi Mounting Plate

This lasered assembly consists of a pair of acrylic plates, one for the Raspberry Pi microcomputer and the other to protect the electronics from damage (see Figure 7.16). It uses aluminum standoffs to connect the two plates together and mounts the assembly to the robot using normal LEGO connector pegs.

FIGURE 7.16 BrickPi mounting plates protect the Raspberry Pi from damage. Credit: Dexter Industries.

3D Printer

The counterpoint to the "subtractive" CNC mill is the "additive" technology of the 3D printer, which begins with an empty bed and adds layer after layer of plastic filament until the shape is formed. The resulting "prints" aren't as strong as solid plastic, but they hold up pretty well. Figure 7.17 shows a part being printed on my 3D printer, a Makeblock Constructor (makeblock.cc) that I built out of a kit.

FIGURE 7.17 My 3D printer outputs a design.

GoPro Camera Mount

This project, shown in Figure 7.18, is the classic example of hacking LEGO bricks: making something that not only LEGO doesn't offer, but I guarantee they never will. If you want to mate your GoPro to a Mindstorms robot, printing a part like this is the best way. Designed by Thingiverse user Chris Magno, the part may be downloaded at http://www.thingiverse.com/thing:13518.

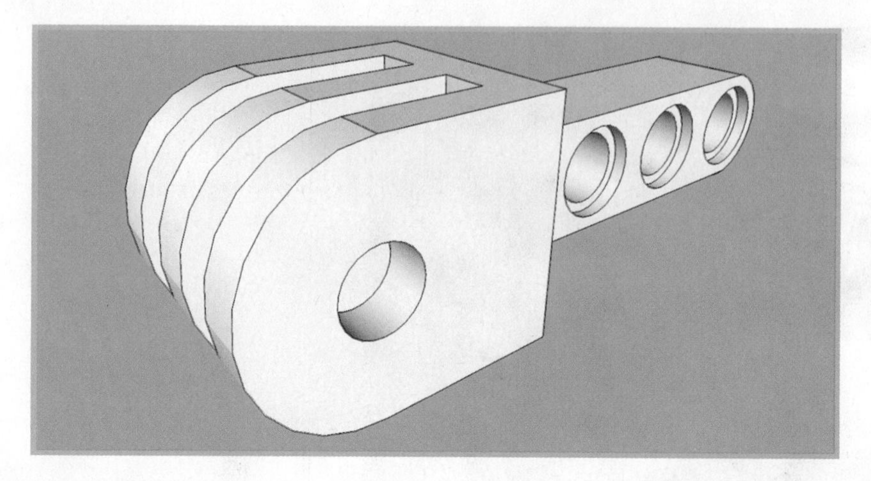

FIGURE 7.18 This GoPro camera mount connects to Mindstorms beams. Credit: Chris Magno (Creative Commons).

Universal Connector Kit

I mentioned that one of the most appealing aspects of designing your own parts is using the technology to mate two dissimilar systems: for example, LEGO and Arduino. The Universal Connector Kit consists of mounting adapters that combine dissimilar building sets. There are 80 total connectors, some of which accommodate rather improbable combinations, such as mating ZOOB to Tinkertoy. The part shown in Figure 7.19 mates Lincoln Logs to LEGO.

You can download individual connectors at Thingiverse, http://www.thingiverse.com/uck/designs or elsewhere on the Web at http://fffff.at/free-universal-construction-kit.

FIGURE 7.19 The Universal Connector Kit mates everything with everything. Credit fffff.at

80/20 Interface

One of the toughest building sets out there is 80/20, which offers massively strong and durable metal parts. As you might expect, folks are forever designing 3D-printable attachments for 80/20, and the LEGO connector plate shown in Figure 7.20 is an example. Thingiverse user Shawn Steele designed the part because he was building a full-sized LEGO R2-D2 robot, and it had an 80/20 chassis with LEGO plates and bricks attached to it. You can download Shawn design's here: http://www.thingiverse.com/thing:347456. You can buy 80/20 at 8020.net.

FIGURE 7.20 This connector plate attaches LEGO to 80/20. Credit: Shawn Steele.

Servo Mount

German Thingiverse user erdinger designed a mounting block for a standard 40mmx20mm servo, and it's equipped with Mindstorms-compatible mounting holes (see Figure 7.21). If you're not going to use Mindstorms' servomotors, the next best thing is to print a mounting block such as this one: http://www.thingiverse.com/thing:7535.

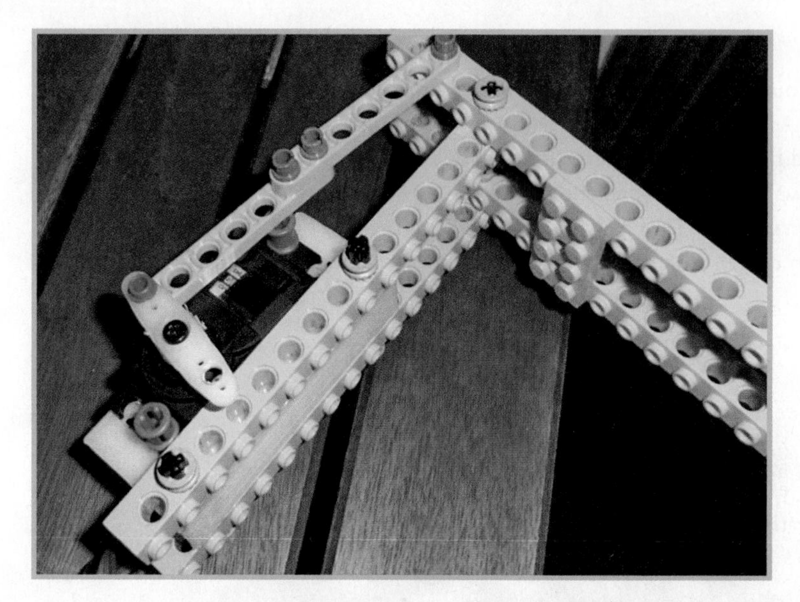

FIGURE 7.21 Print this block to secure a servo to your Mindstorms robot. Credit: erdinger (Creative Commons).

Mechanum Wheel

Mechanum or omni wheels work the same as a normal wheel, but they also have small wheels around the edge, allowing the robot to move perpendicular to the wheel's axis. Typically this is accomplished by giving the robot four such wheels on two axles 90 degrees from each other. When you want the robot to go forward, the left and right mechanum wheels turn. When you want the robot to move to the left or right, the front and back wheels are turned. Dutch Thingiverse user projunk designed the wheels shown in Figure 7.22, and you can download it here: http://www.thingiverse.com/thing:681259.

FIGURE 7.22 Small wheels around the mechanum wheel's rim allow the robot to go sideways. Credit: projunk.

Tip: Parametric 3D Models

Parametric models are ones where the measurements and configuration of the model can be adjusted numerically, either by manually modifying the build instructions or by using a software tool that does it for you. Steve Medwin's Customizable Technic Hub (http://www.thingiverse.com/thing:703993) serves as a great example of how parametric 3D models work.

I told you CNC tools get simple instructions from the computer, which interprets a vector graphic to figure out what direction to send the toolhead and for how long. However, the advantage of this numerical system is that you can modify the numbers to make a new model.

In the case of the Technic hub project, you can take the design into Thingiverse's Customizer application, type in different options, and get a preview of how it will look, as shown in Figure 7.23. Steve's design lets you print a disc with anywhere from 8 to 36 holes and either a cross-shaped mounting hole or one or more larger holes in the center.

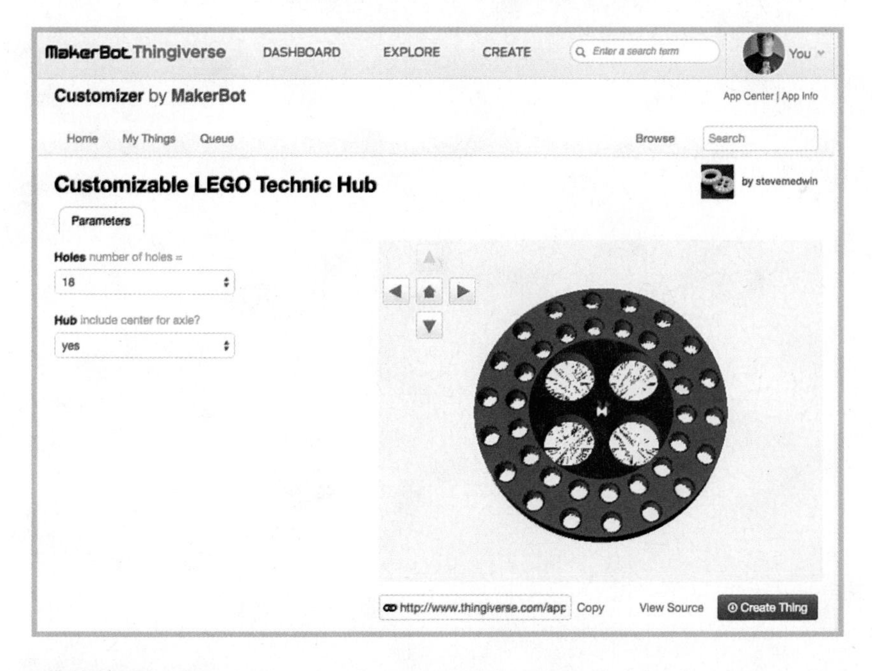

FIGURE 7.23 Don't like the Customizable Technic Hub? Just remix it.
Credit: Steve Medwin.

Summary

This chapter covered some of the many options for adding onto your Mindstorms set's beam set. Connectors, plates, and beams all can be output with the help of computer-controlled tools. In Chapter 8, "Project: Ball Contraption," you have the option to put these skills to the test. First you build a Ball Contraption, a motorized course that runs Mindstorms balls around a track. Then you create some custom parts that make the contraption work even better.

Project: Ball Contraption

Ball contraptions are popular projects in the LEGO community. They are machines that continually cycle LEGO balls through a series of conveyors, chutes, and other obstacles simply for the fun of it. This chapter's project is just such a contraption, and you can see it in Figure 8.1.

After you're done building and programming the robot out of Mindstorms parts, you put to use the skills you learned in Chapter 7, "Hacking LEGO III: Create Your Own LEGO Parts," by designing and manufacturing two different add-on parts, one using a 3D printer and one making use of a laser cutter. These not only help you make your contraption better, but also give you the skills to build future parts.

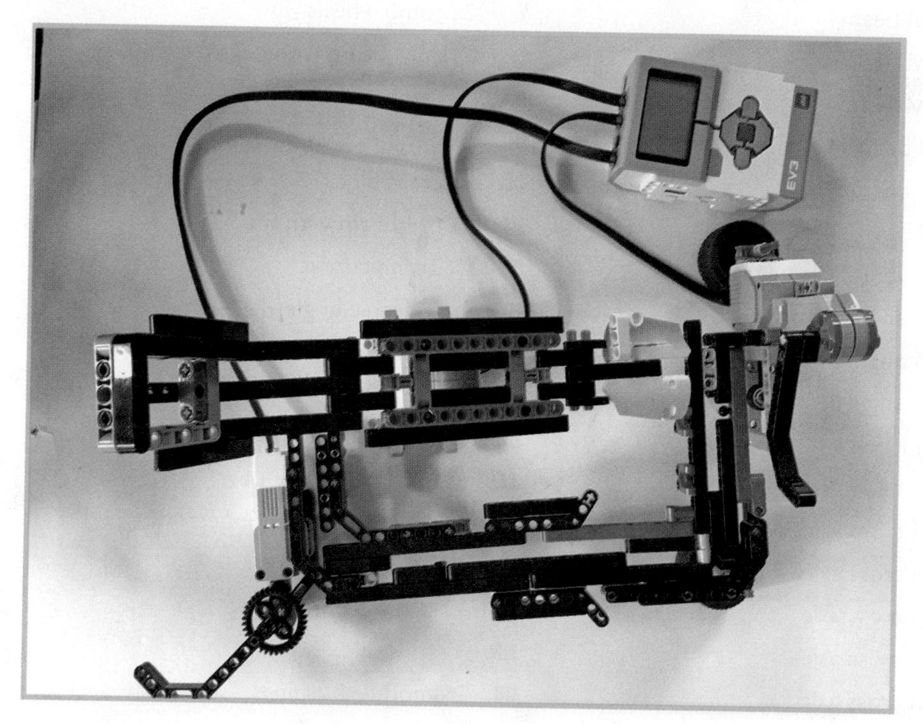

FIGURE 8.1 The ball contraption moves a ball around a course. But how could you make it better?

Building the Contraption

The contraption consists of two main parts: I call one the "teeter-totter" because it kind of looks like one, and the other is the "course" because most of it is the track along which the ball rolls.

Limited as I was by the constraint of using only a single EV3 set to build the contraption, I found my beam reserves stretched to the limit. Some of the connections, therefore, weren't the way I would have done them if I had unlimited parts! This only serves to underscore the need for occasionally designing and manufacturing your own beams, plates, and connectors.

Parts List

You need the following parts from the LEGO Mindstorms EV3 set. You can find each part in Figure 8.2.

A. 1x EV3 Intelligent brick
B. 2x medium motors
C. 1x small motor
D. 3x 15M beams
E. 4x 13M beams
F. 4x 11M beams
G. 4x 9M beams
H. 7x 7M beams
I. 7x 5M beams
J. 10x double-angle beams
K. 5x 3x7 angle beams
L. 4x 3x5 angle beams
M. 6x 2x4 angle beams
N. 2x 5x11 beam frames
O. 2x 5x7 beam frames
P. 2x 8M cross axles with end stop
Q. 2x 7M cross axles
R. 1x 6M cross axle
S. 6x 5M cross axles
T. 1x 3M cross axle
U. 1x Z12 conical gear
V. 1x Z20 bevel gear
W. 1x worm gear
X. 1x Z36 gear
Y. 2x comb wheels
Z. 11x 3M beam with pegs
AA. 2x catch with cross hole
BB. 6x double cross blocks
CC. 1x cross block
DD. 1x 3M beam with fork
EE. 2x Technic forks
FF. 1x 0-degree angle element
GG. 1x tube
HH. 2X rims and tires
II. 1X small rim and tire
JJ. 12x 3M connector pegs
KK. 1x 3M no-friction connector peg
LL. 64x connector pegs
MM. 27x cross connectors
NN. 1x towball peg
OO. 2x module bushings

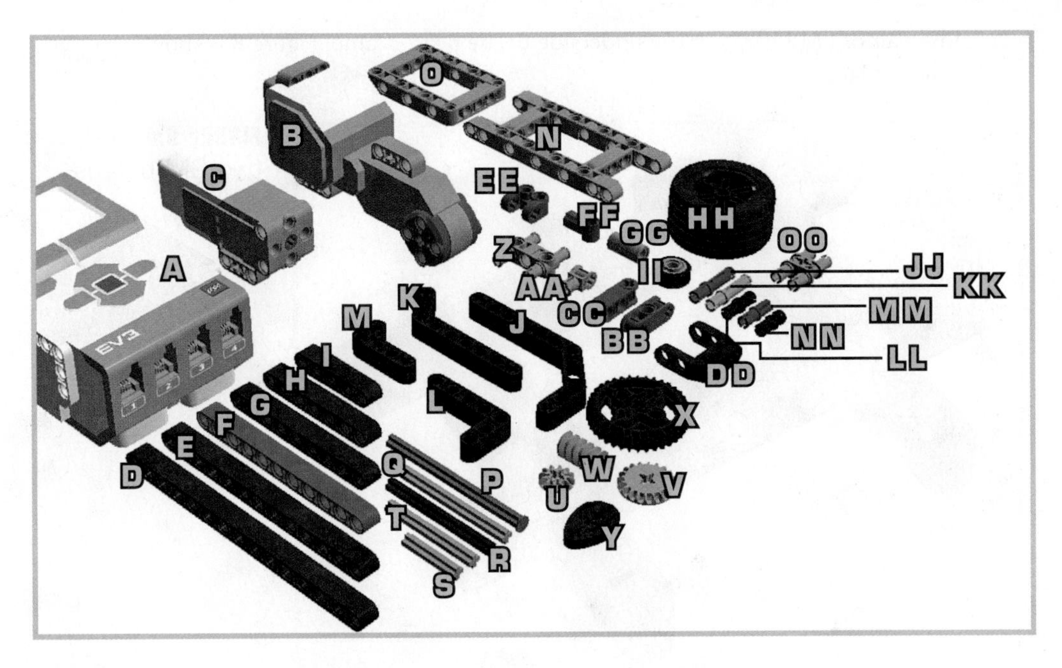

FIGURE 8.2 You need these parts to build the contraption.

Steps

Let's begin building! Once you have your parts together (as usual, they can all be found in the EV3 set) you can begin building your contraption.

1. Grab the 5x11 beam frame and add eight connector pegs as shown in Figure 8.3.

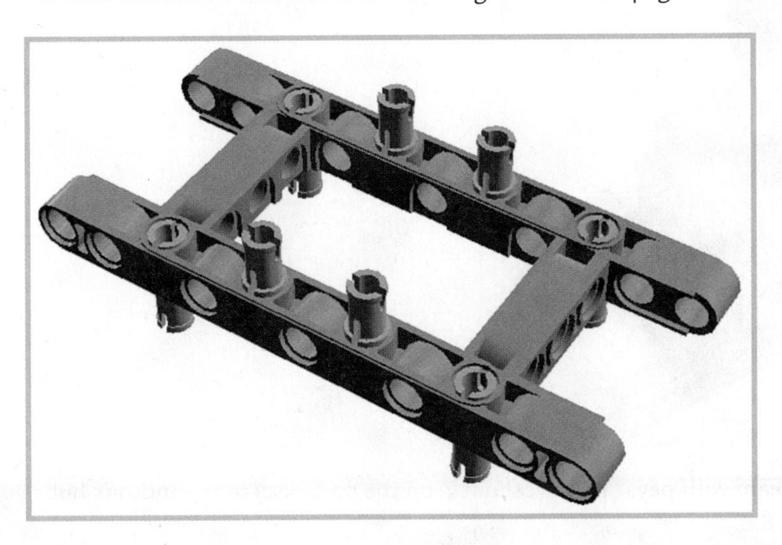

FIGURE 8.3 Insert 8 pegs into the beam frame.

2. Add a pair of 9M beams to the underside of the beam frame. Figure 8.4 shows the beams in place.

FIGURE 8.4 Add two 9M beams.

3. Attach a motor to the free pegs on the beam frame (see Figure 8.5).

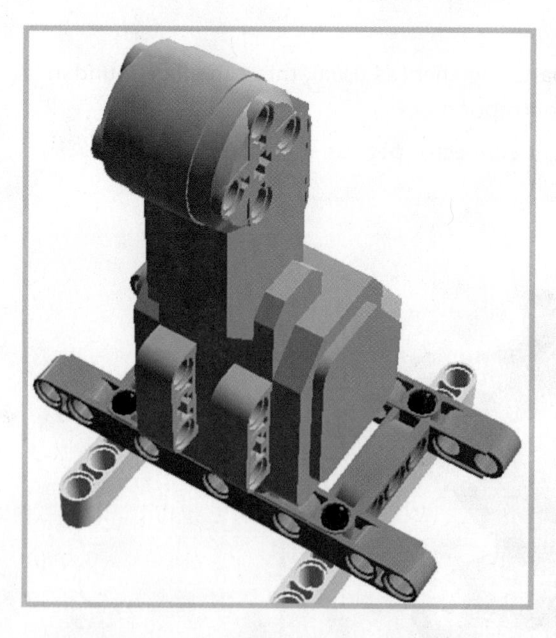

FIGURE 8.5 Attach a motor to the beam frame.

4. Attach four "3M beam with pegs" parts (callout Z on the parts list) to the motor's hub as shown in Figure 8.6.

FIGURE 8.6 Four "3M beam with pegs" parts attach to the hub.

5. Connect another 5x11 beam frame to the 3M beams with pegs. Figure 8.7 shows how it should look.

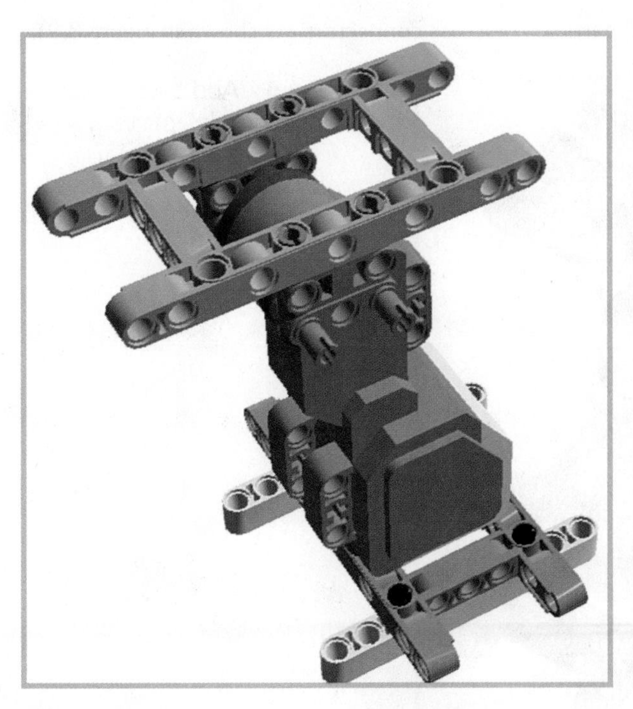

FIGURE 8.7 Another beam frame becomes the basic platform of the teeter-totter.

6. Insert a pair of cross connectors to the ends of the beam frame (see Figure 8.8).

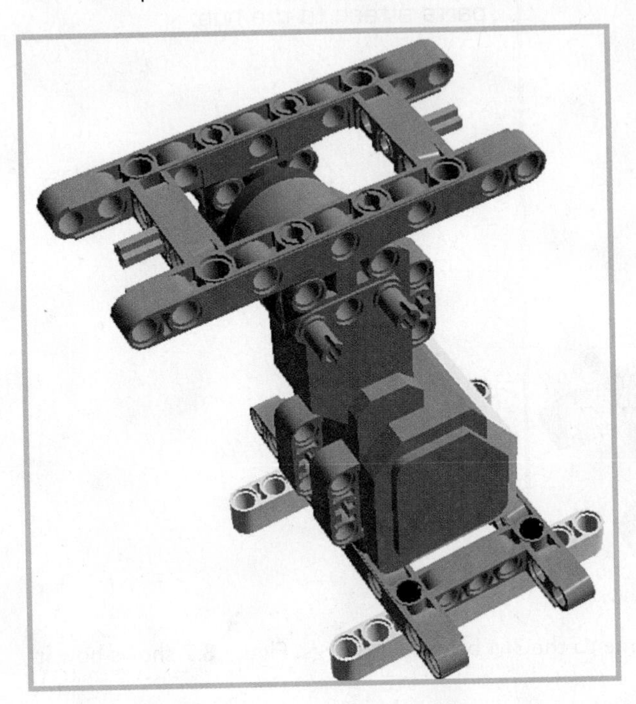

FIGURE 8.8 Add two cross connectors.

7. Add two "catch with cross hole" parts to the cross connectors (as shown in Figure 8.9).

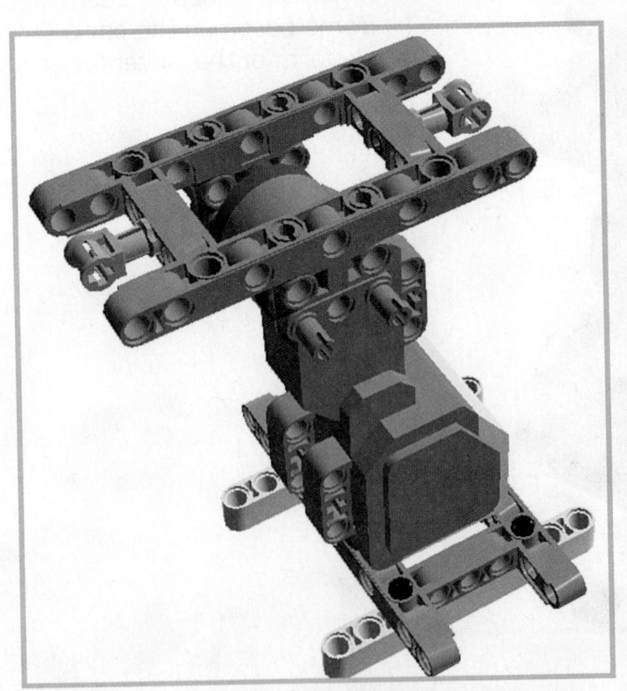

FIGURE 8.9 Add two "catch with cross hole" parts.

8. Connect a pair of 2x4 angle beams with the help of a 7M cross axle. Figure 8.10 shows how it should look.

FIGURE 8.10 Secure a pair of angle beams with a cross axle.

9. Insert a 3x7 angle beam between the 2x4 beams, and secure it with a 5M cross axle, as shown in Figure 8.11. Also add a pair of connector pegs while you're there.

FIGURE 8.11 Add an angle beam and secure with a cross axle.

10. Add two more 3x7 angle beams as shown in Figure 8.12.

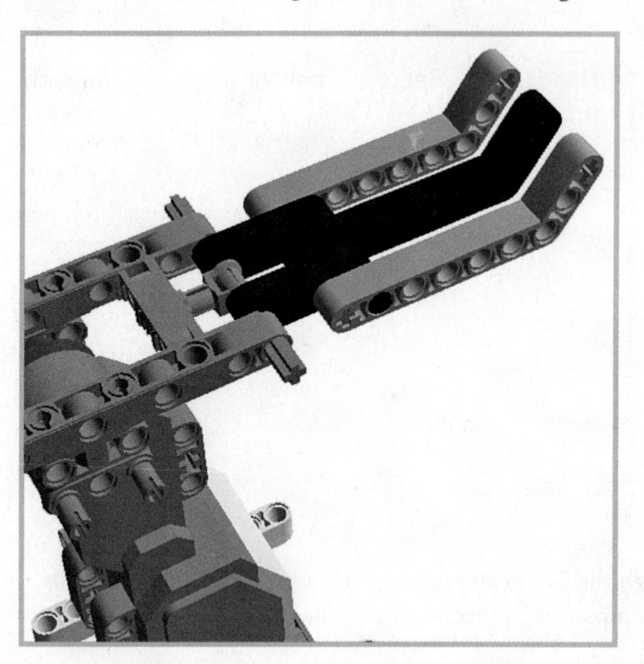

FIGURE 8.12 Attach two more angle beams.

11. Attach a connector peg and a cross connector to one of the angle beams (see Figure 8.13).

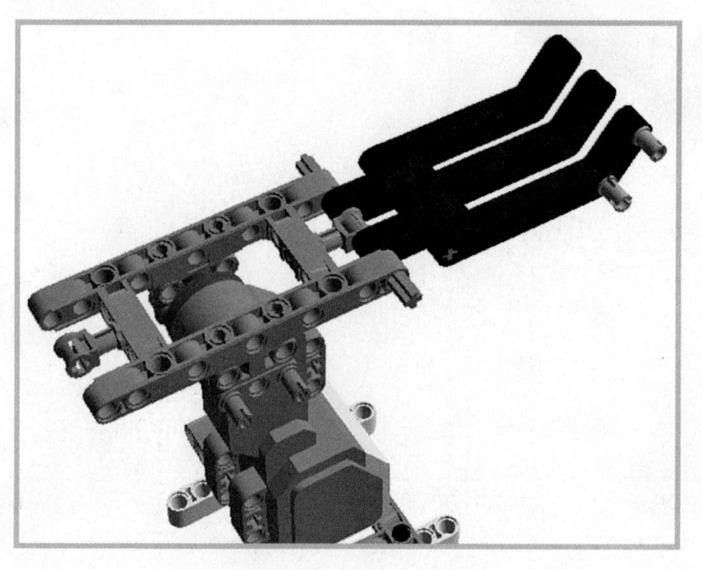

FIGURE 8.13 Add two pegs to one of the beams.

12. Add a 7M beam to the structure, plus a couple of connector pegs. Figure 8.14 shows how it should look.

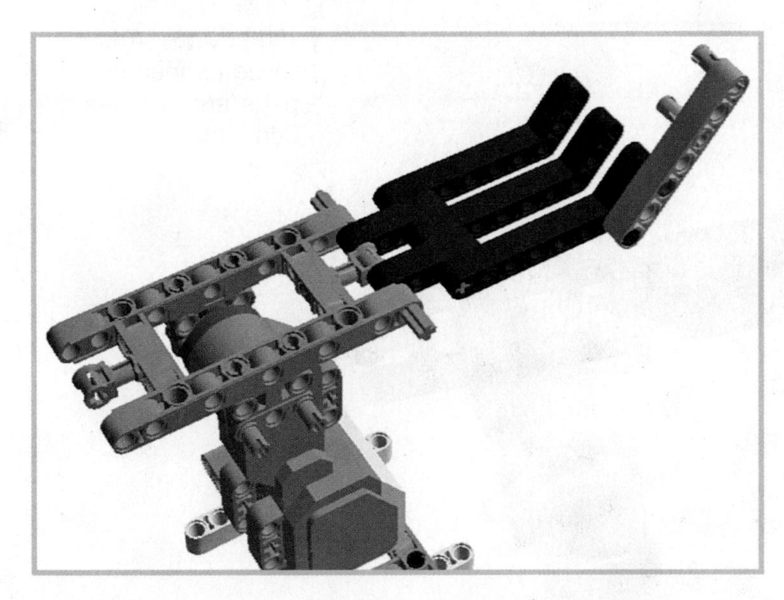

FIGURE 8.14 Another 7M beam is added, along with a couple of pegs.

13. Attach a 5x7 beam frame to the 7M beam, as shown in Figure 8.15.

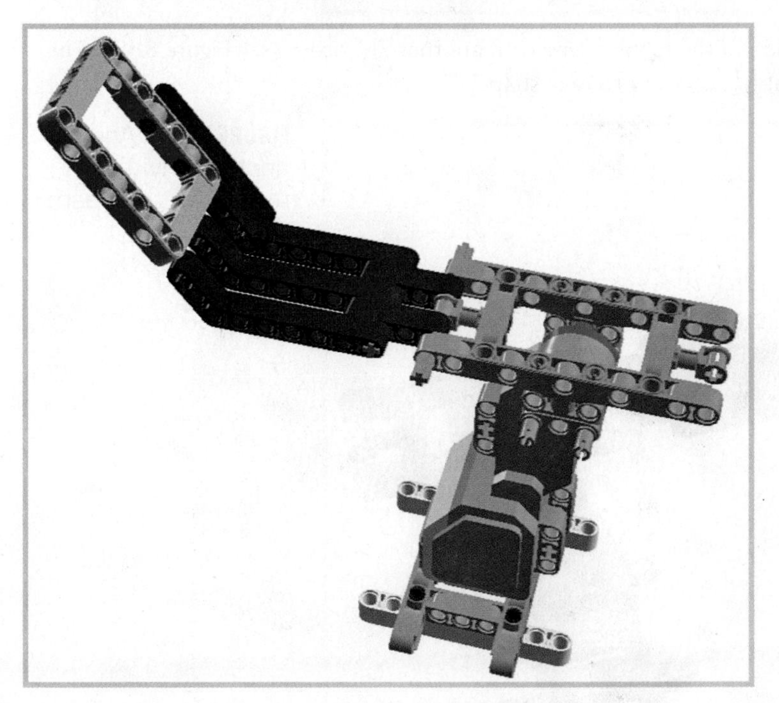

FIGURE 8.15 Add a beam frame to the 7M beam.

14. Insert three regular connector pegs and one cross connector. Figure 8.16 shows how it should look.

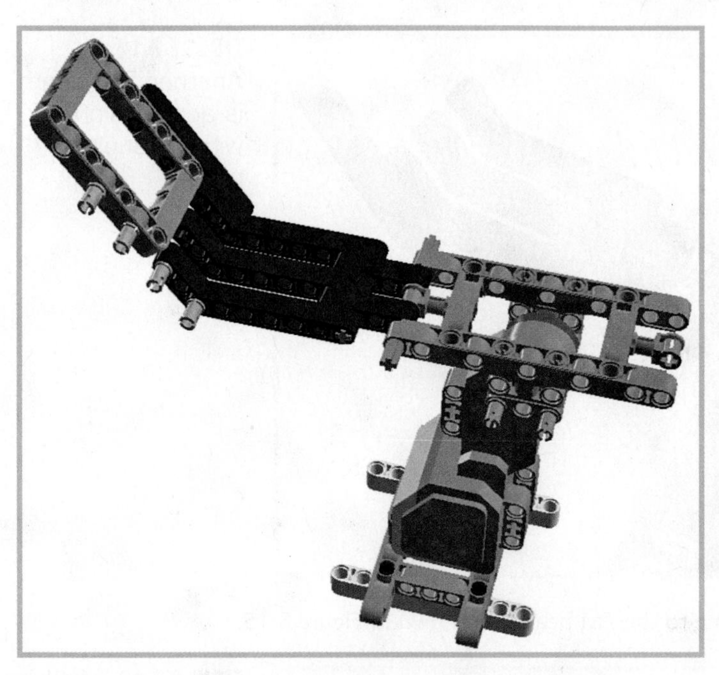

FIGURE 8.16 Add three connector pegs and a cross connector.

15. Secure the other side of the beam frame with another 7M beam (see Figure 8.17). The teeter-totter assembly is starting to take shape!

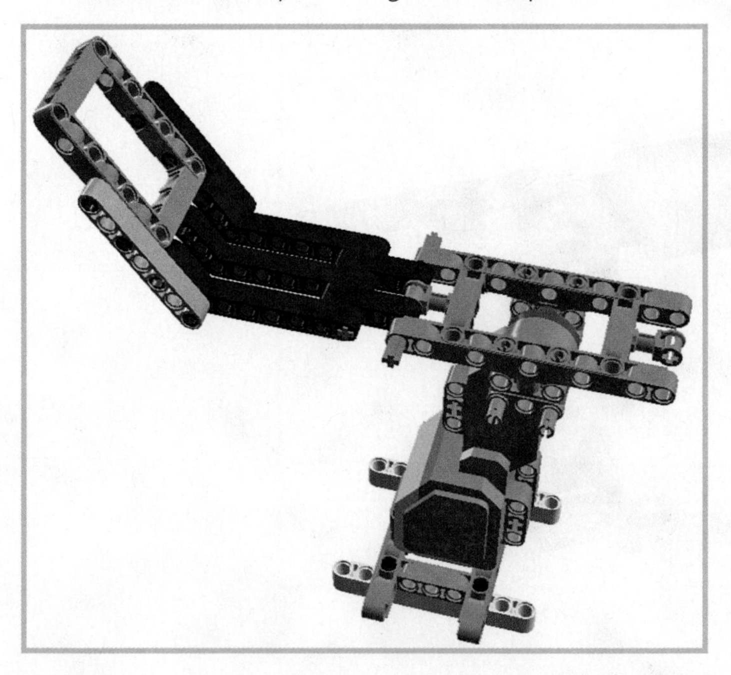

FIGURE 8.17 Add another 7M beam to secure the beam frame.

16. Insert a pair of double cross blocks and four cross-connectors to secure them. Figure 8.18 shows how it should look.

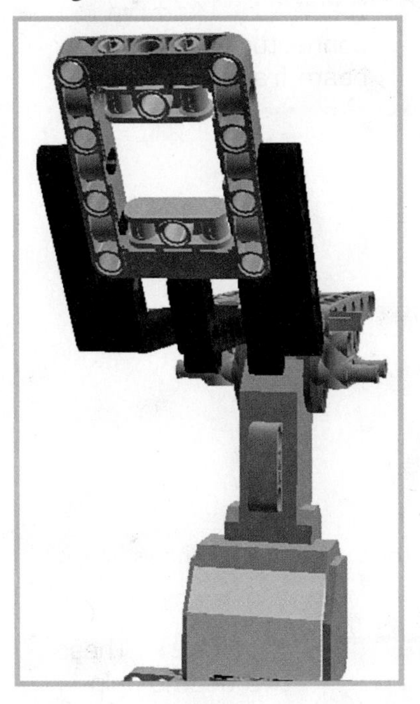

FIGURE 8.18 Add two double cross blocks.

17. Add a 5M beam to the double cross blocks with the help of a pair of connector pegs (see Figure 8.19).

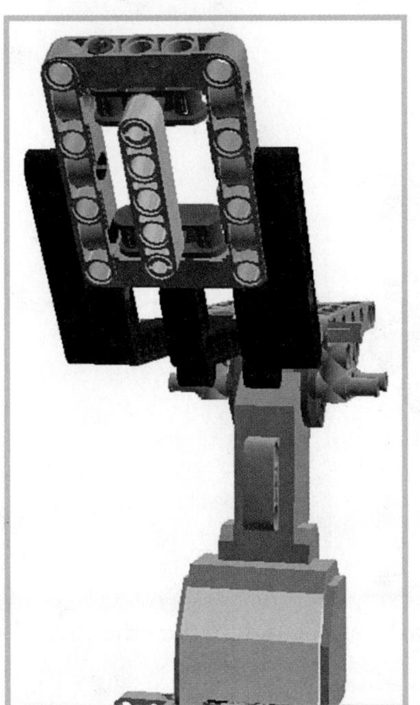

FIGURE 8.19 Attach a 5M beam to the double cross blocks.

18. Insert five connector pegs to the 5x7 beam frame, as shown in Figure 8.20.

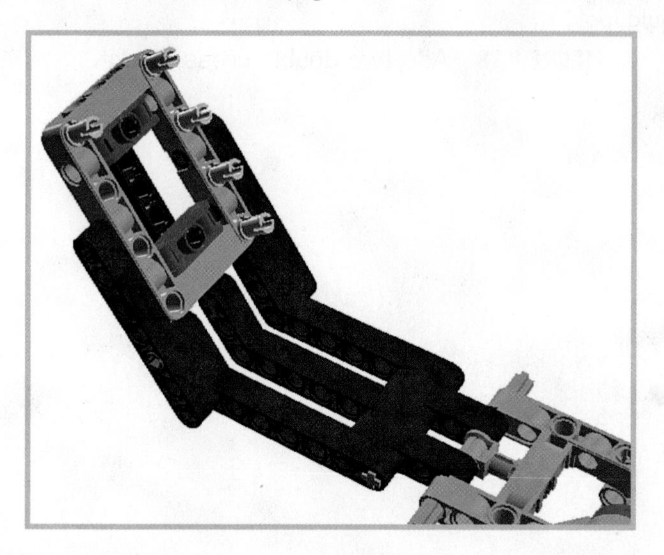

FIGURE 8.20 Add five connector pegs to the beam frame.

19. Add a 3x5 angle beam and a 5M beam to the teeter-totter. Figure 8.21 shows how it should look.

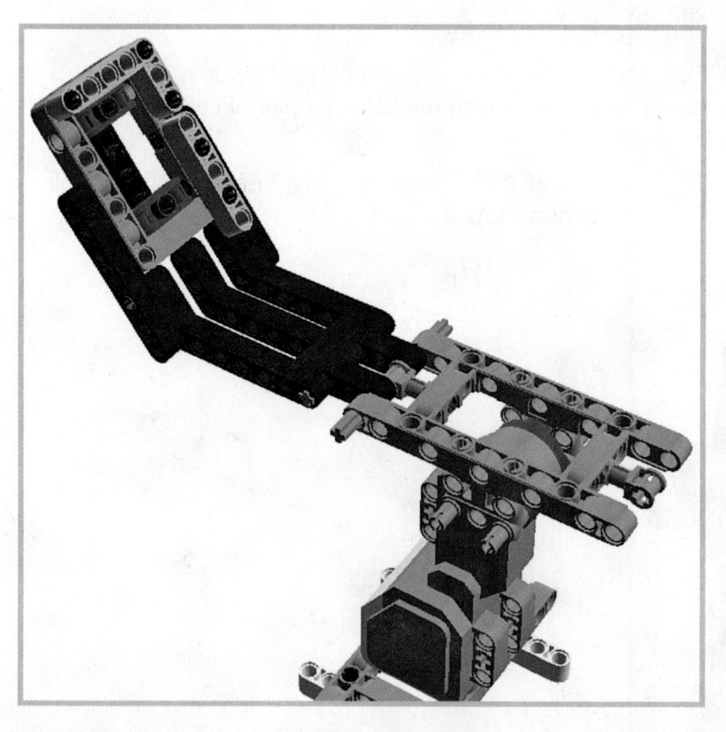

FIGURE 8.21 These two beams help reinforce the contraption.

20. Attach another pair of 2x4 angle beams using a 7M axle, as you did on the other side (see Figure 8.22).

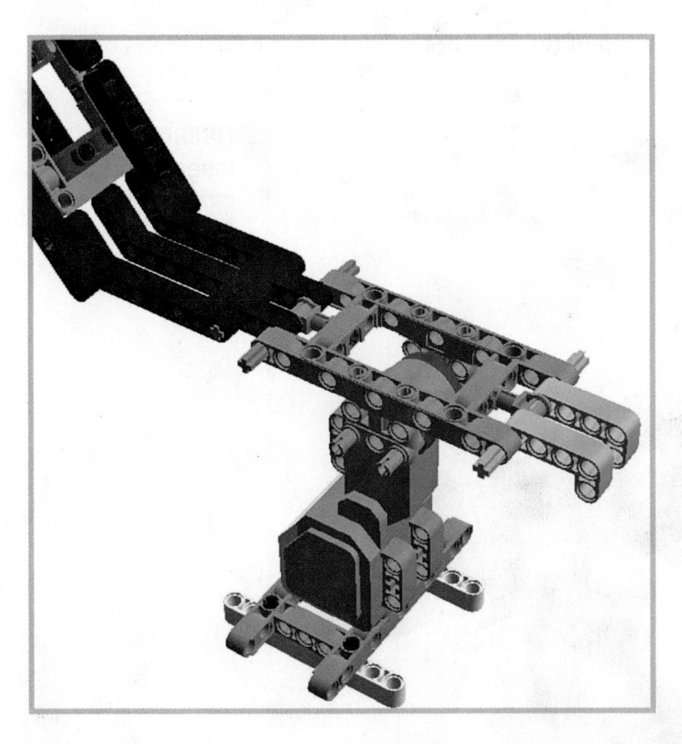

FIGURE 8.22 Add two more angle beams.

21. Use 3M connector pegs (the long blue ones) to attach a pair of 5M beams to the middle of the 5x11 beam frame, as shown in Figure 8.23.

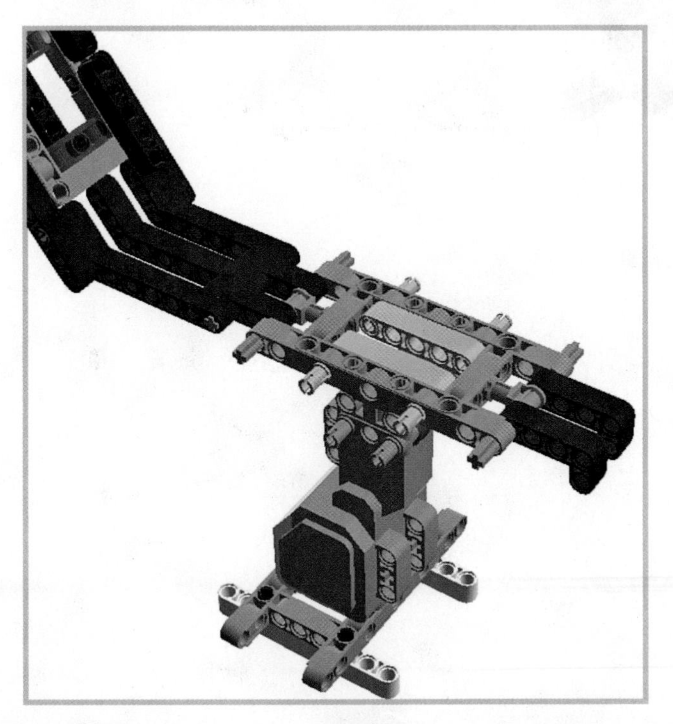

FIGURE 8.23 Attach a pair of 5M beams with the help of 3M pegs.

22. Add four connector pegs to the top of the 5x11 beam frame. Figure 8.24 shows how it should look.

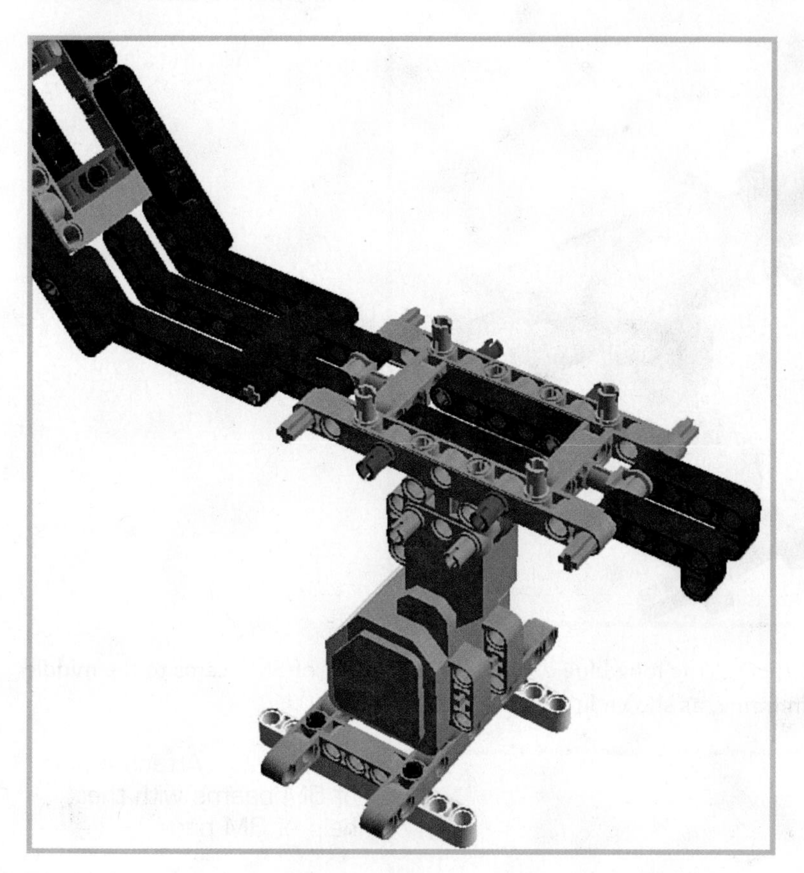

FIGURE 8.24
Insert four pegs into the beam frame.

23. Add a pair of red 11M beams and a pair of black 13M beams to the teeter-totter, as shown in Figure 8.25.

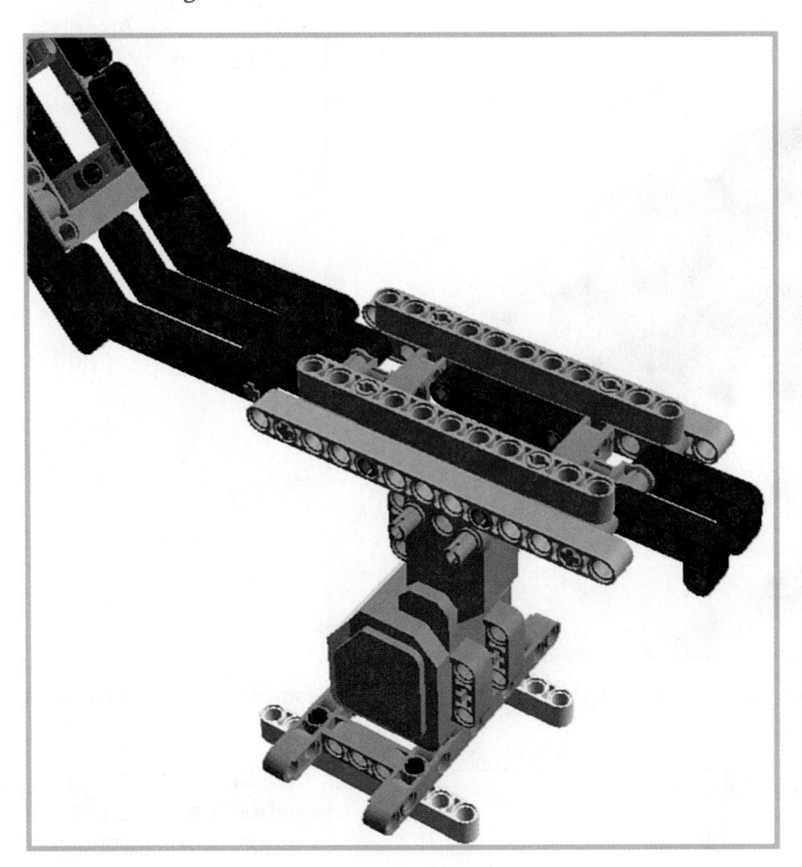

FIGURE 8.25
Attach a bunch of beams to reinforce the teeter-totter.

24. Insert a 7M beam between the two 2x4 angle beams, then secure with a pair of 5M axles (see Figure 8.26).

FIGURE 8.26 Attach a 7M beam.

25. Secure the ends of the cross axles with a pair of comb wheels (sometimes called cams) shown in Figure 8.27.

FIGURE 8.27 Comb wheels secure the ends of the cross axles.

26. In the final step of the teeter-totter, add a pair of 3x7 panels to the 7M beam (as shown in Figure 8.28).

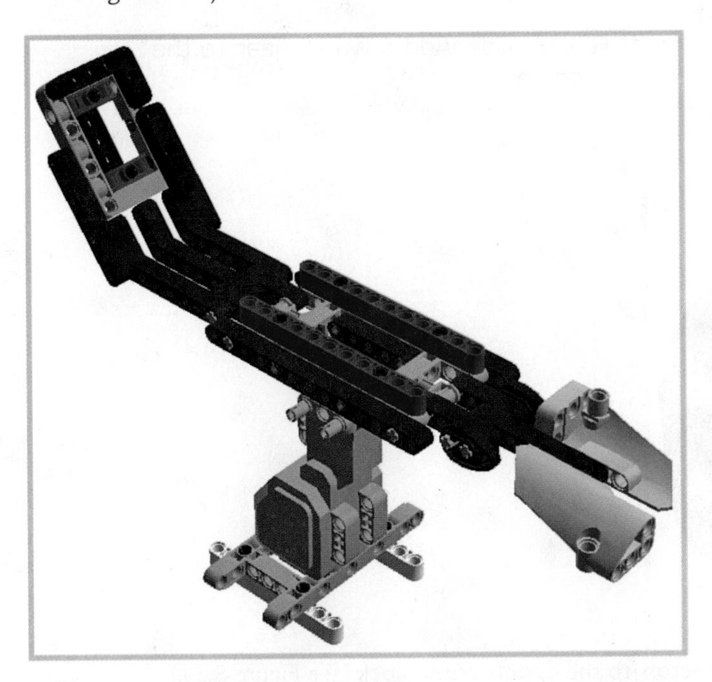

FIGURE 8.28 Add the panels.

27. Let's begin working on the ball course! Begin with a tire and rim and add an 8M beam with end stop, securing with a bushing (shown in Figure 8.29).

FIGURE 8.29 The ball course begins with a tire and rim.

28. Add a worm gear, a double cross block, and another bushing to the axle. Figure 8.30 shows how it should look.

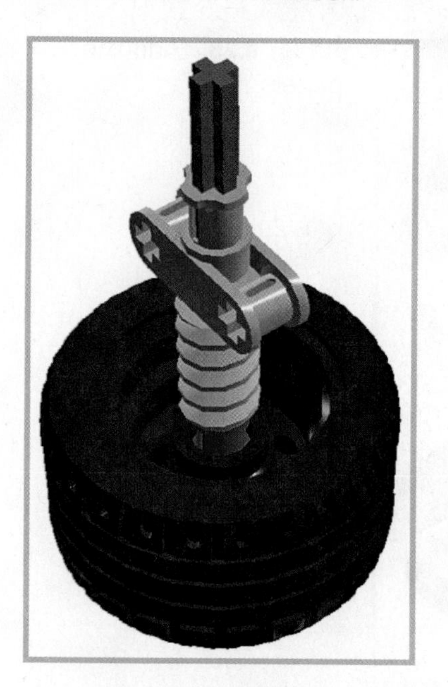

FIGURE 8.30 Add a worm gear to the cross axle.

29. Insert a pair of cross connectors to the double cross block (see Figure 8.31).

FIGURE 8.31 Insert a pair of cross connectors.

30. Add a 3M beam with pegs to the round ends of the cross connectors (see Figure 8.32).

FIGURE 8.32 Add a 3M beam with pegs.

31. Attach the assembly to a motor, as shown in Figure 8.33.

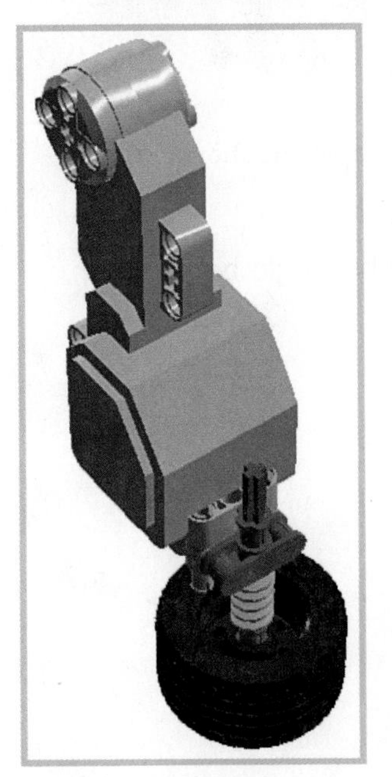

FIGURE 8.33 Add a motor to the wheel assembly.

32. On the motor's hub, attach a 2M beam with a 3M connector peg and a 3M axle. Figure 8.34 shows how it should look.

FIGURE 8.34 Attach a 2M beam, a peg, and a 3M axle.

33. Attach a double-angle beam to the motor using a pair of 5M axles (as shown in Figure 8.35).

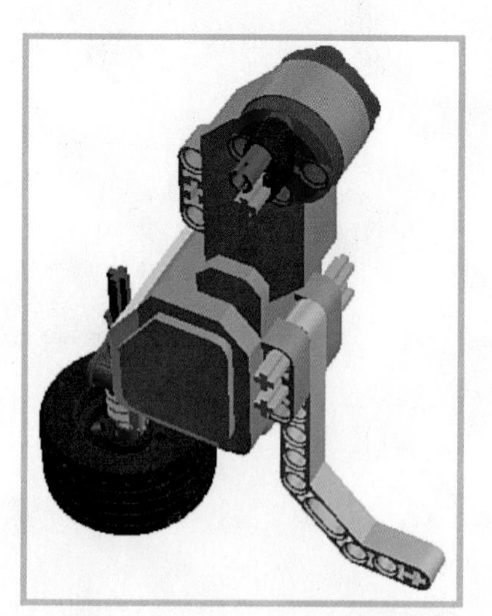

FIGURE 8.35 Add a double-angle beam.

34. Grab three more double-angle beams. Put two on the pair of 5M axles and then attach the third one to the motor's hub. Figure 8.36 shows how it should look.

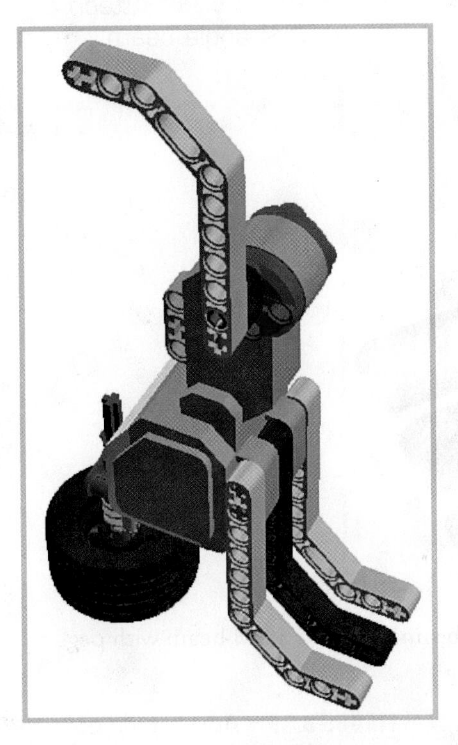

FIGURE 8.36 Add three more double-angle beams.

35. Insert a pair of connector pegs to one of the double-angle beams, as shown in Figure 8.37.

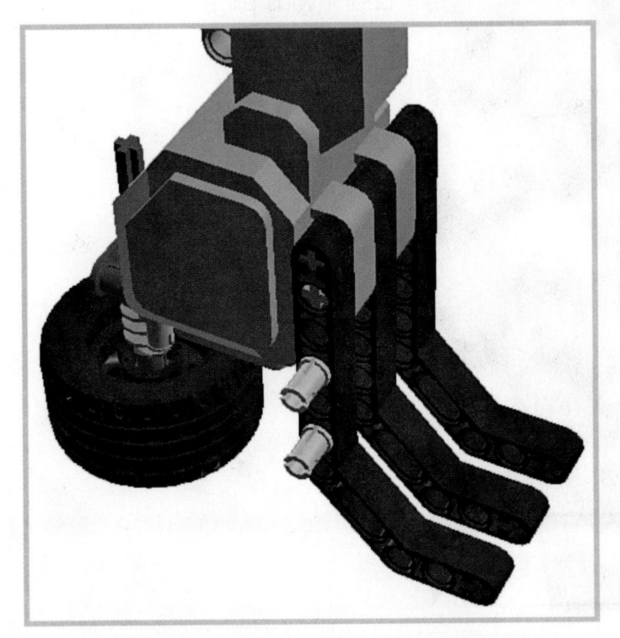

FIGURE 8.37 Insert a pair of pegs to one of the beams.

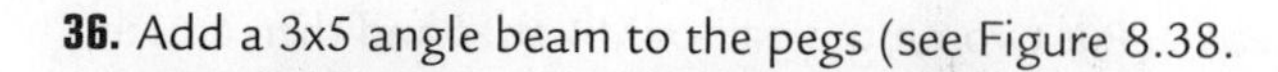

36. Add a 3x5 angle beam to the pegs (see Figure 8.38.

FIGURE 8.38 Attach a 3x5 angle beam.

37. Insert a pair of connector pegs to the 3x5 angle beam, and add a 3M beam with pegs as well (see Figure 8.39).

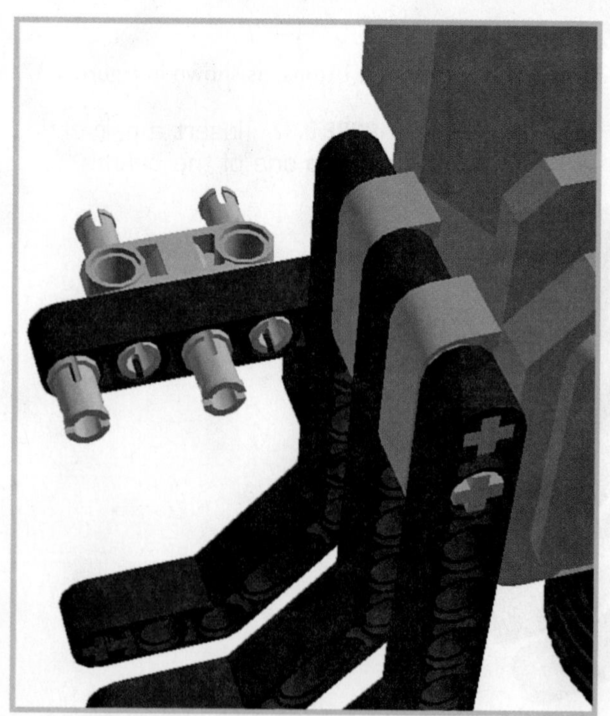

FIGURE 8.39 Add a pair of connector pegs and a 3M beam with pegs.

38. Attach a red 11M beam to the pegs, inserting four more pegs while you're at it. Figure 8.40 shows the current state of the course assembly.

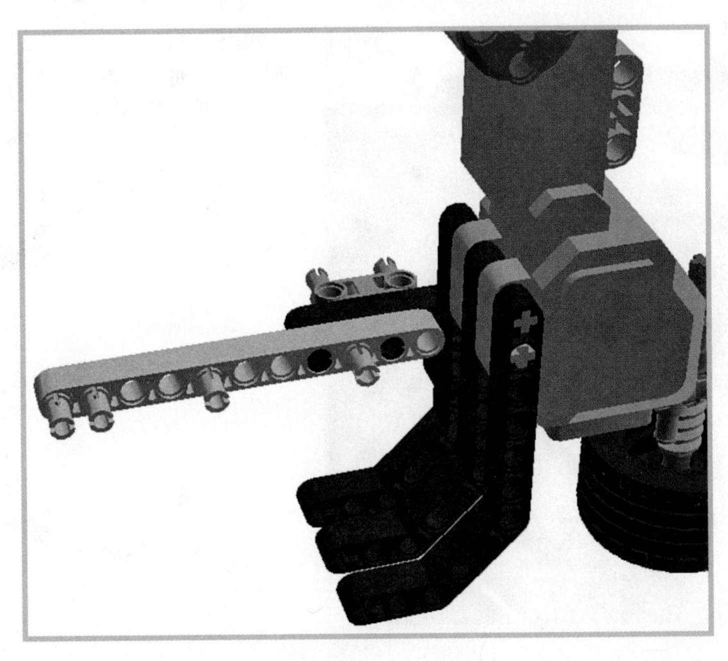

FIGURE 8.40 Attach a beam and four additional pegs.

39. Add a 5x7 beam frame to the 11M beam. This frame serves as a bulwark against the ball escaping after the teeter-totter flips it (see Figure 8.41).

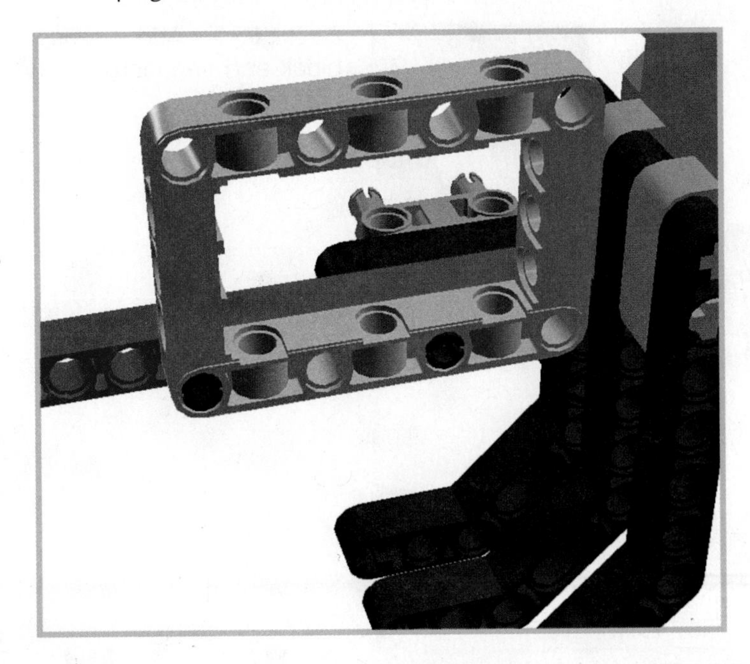

FIGURE 8.41 The beam frame helps keep the ball on course.

40. Add a 3M no-friction peg (they're beige) and one of the small wheels that came with the set. Figure 8.42 shows how it should look.

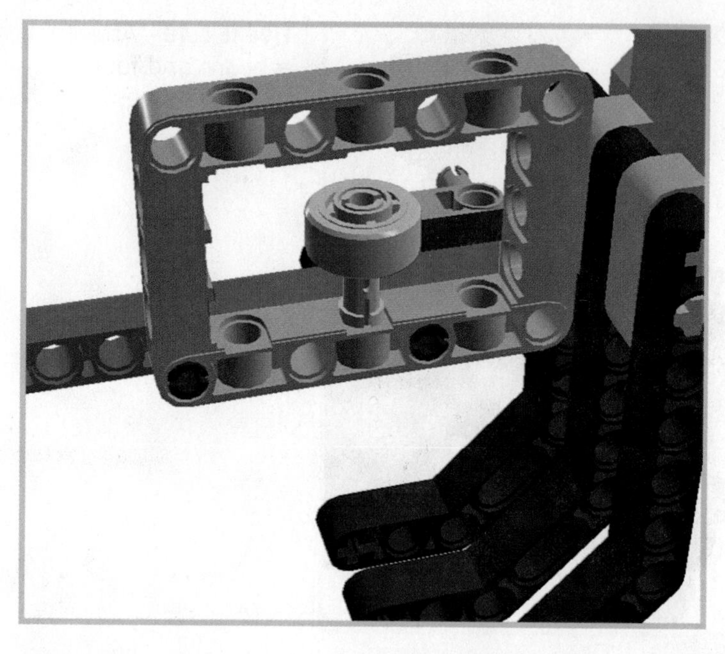

FIGURE 8.42 A small wheel helps keep the ball from escaping.

41. Add a cross block and a connector peg to the end of the red 11M beam, as shown in Figure 8.43.

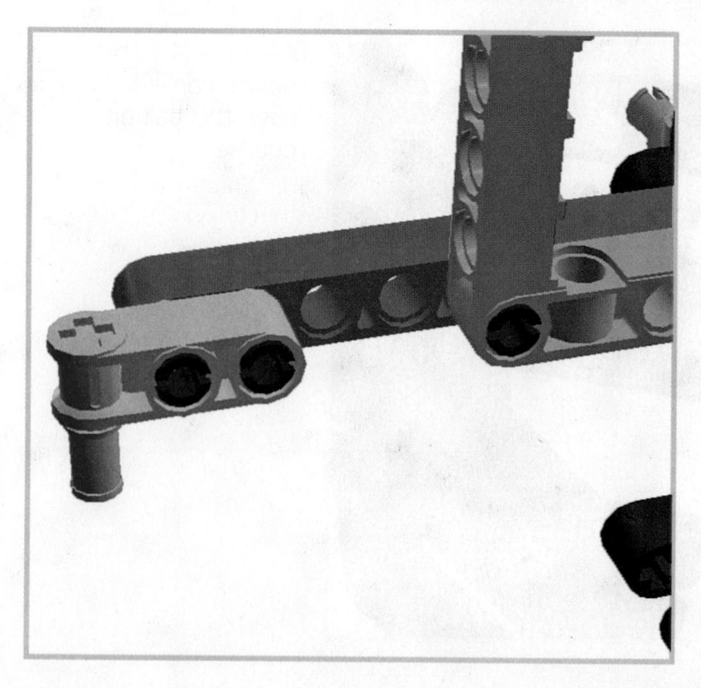

FIGURE 8.43 Add a cross block and connector peg.

42. Connect a 5M beam to the cross block you just added (see Figure 8.44).

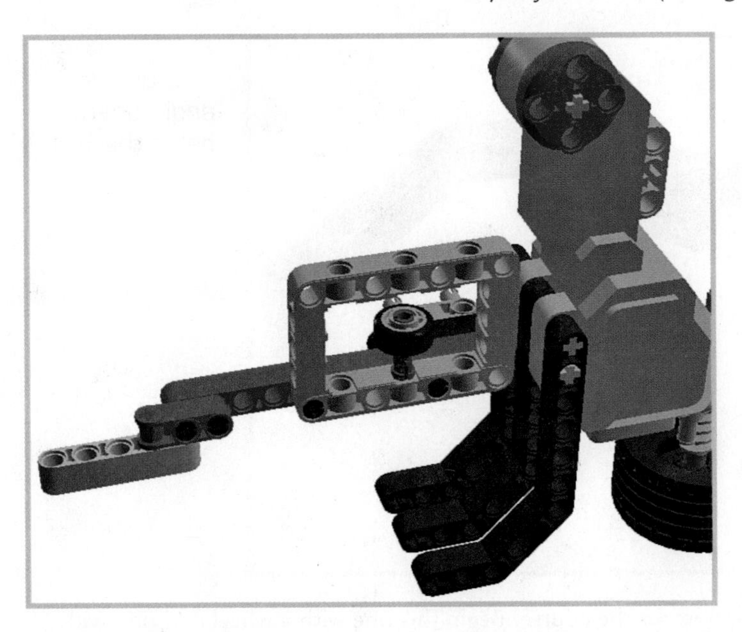

FIGURE 8.44 Attach a 5M beam.

43. Insert three connector pegs, a cross connector, and a towball peg to the assembly (see Figure 8.45).

FIGURE 8.45 Add a bunch of pegs!

44. Help the ball turn the corner with a double-angle beam, as shown in Figure 8.46.

FIGURE 8.46 This double-angle beam helps the ball stay on course.

45. Let's work on a new support for the course. Begin this one with a wheel and rim, with an 8M axle with end stop and a bushing to secure it (see Figure 8.47).

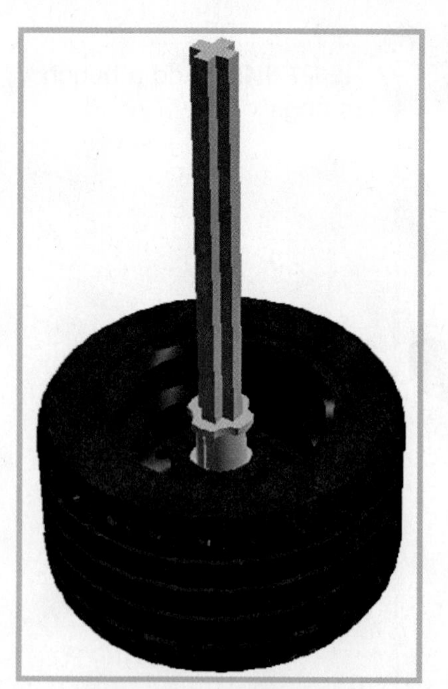

FIGURE 8.47 Another wheel and rim serve to support the course.

46. Slide a worm gear onto the cross axle, as shown in Figure 8.48.

FIGURE 8.48 Slide a worm gear onto the axle.

47. Install the support to the double-angle beam. Figure 8.49 shows how it should look.

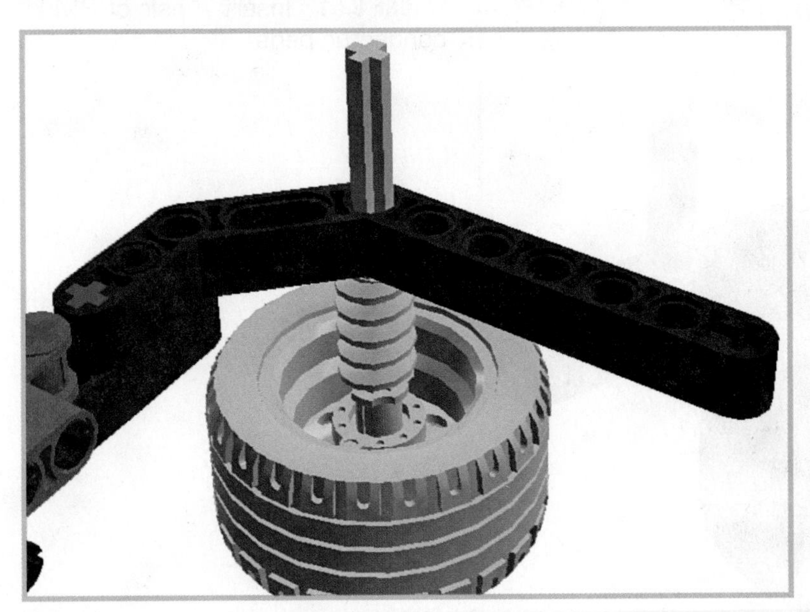

FIGURE 8.49 Attach the support to the double-angle beam by sliding it in from below.

48. Add a bushing to secure the support (see Figure 8.50).

FIGURE 8.50 A bushing helps to secure the support.

49. Insert a pair of 3M connector pegs to the double-angle beam. The course continues to take shape, as shown in Figure 8.51.

FIGURE 8.51 Insert a pair of 3M connector pegs.

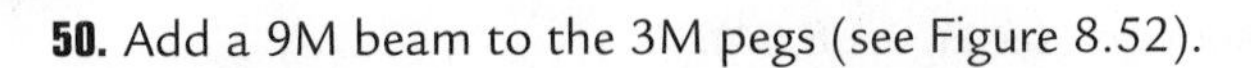

50. Add a 9M beam to the 3M pegs (see Figure 8.52).

FIGURE 8.52 Slide a 9M beam onto the 3M pegs.

51. Add a double cross block to the 9M beam with the help of a couple of cross connectors. While you're at it, add a connector peg, as shown in Figure 8.53.

FIGURE 8.53 Attach a double cross block.

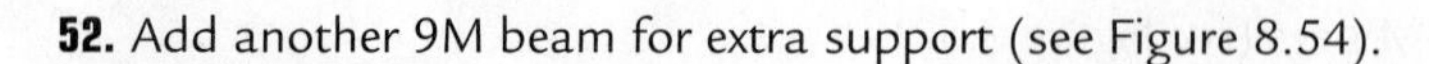

52. Add another 9M beam for extra support (see Figure 8.54).

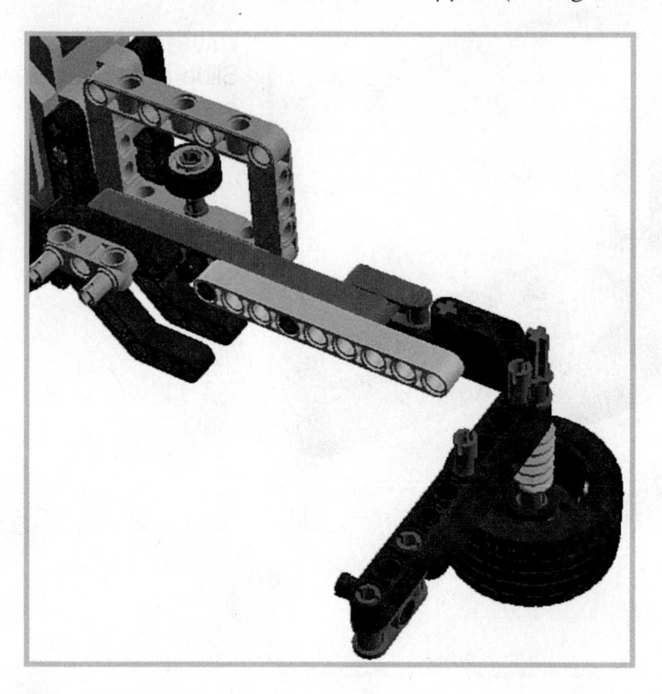

FIGURE 8.54 Add another beam.

53. Attach a 3x5 angle beam to the course with the help of a pair of 3M pegs (shown in Figure 8.55).

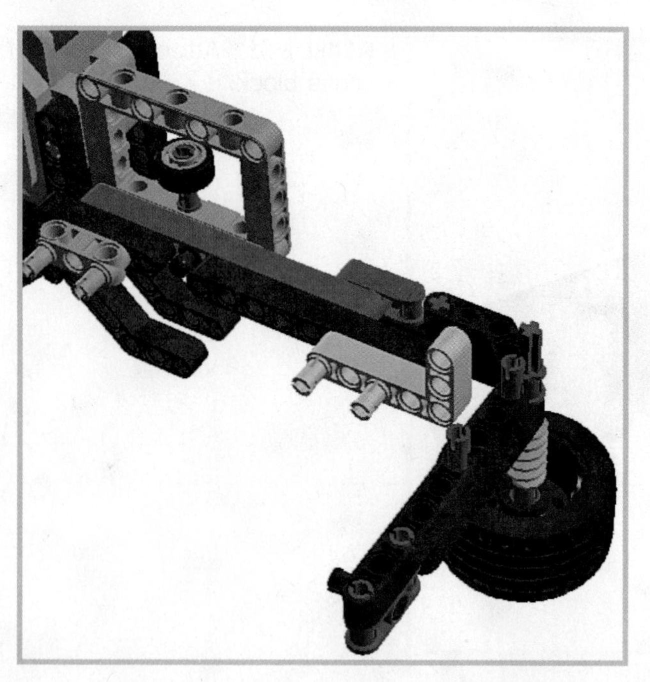

FIGURE 8.55 Add an angle beam.

54. The course continues to take shape with the addition of a 15M beam and a double-angle beam, as shown in Figure 8.56.

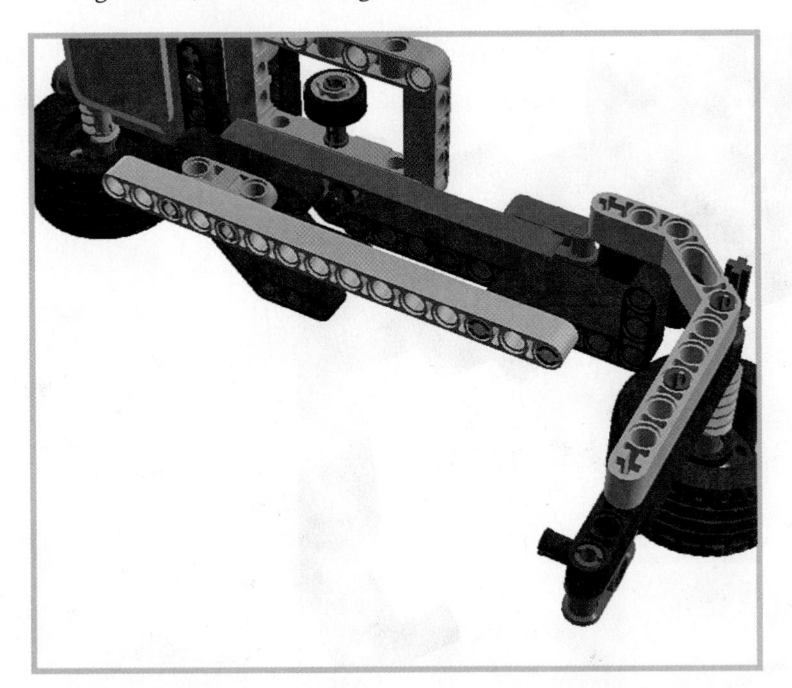

FIGURE 8.56 Attach a 15M beam and a double-angle beam.

55. Insert four regular and one cross connector, as shown in Figure 8.57.

FIGURE 8.57 Add a bunch of pegs!

56. Add some guardrails! Figure 8.58 shows the addition of a pair of cross block forks, which help to keep the ball from falling out.

FIGURE 8.58 The cross block forks help keep the ball from falling off the course.

57. Add a 0-degree angle element with a connector peg as well. Figure 8.59 shows the element in place.

FIGURE 8.59 Add a 0-degree angle element.

58. Attach a red 11M beam and a pair of connector pegs as shown in Figure 8.60.

FIGURE 8.60 Attach a 11M beam.

59. Add a 5M beam to the pegs. Figure 8.61 shows how it should look.

FIGURE 8.61 Add a 5M beam.

60. Attach a 13M beam as shown in Figure 8.62. This is the outer rail of the course.

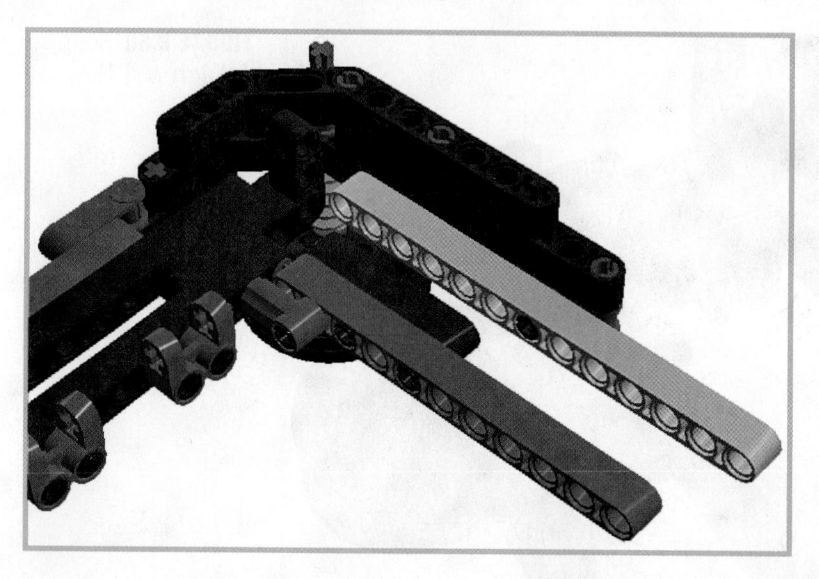

FIGURE 8.62 Attach a 13M beam.

61. Insert a no-friction 3M peg and a pipe element to the 13M beam, as shown in Figure 8.63.

FIGURE 8.63 The pipe element helps keep the ball from falling out.

62. Use a 3M beam with pegs to secure a 7M beam (see Figure 8.64).

FIGURE 8.64 Add a 7M beam.

63. Lengthen the course by adding a 15M beam along with another pair of 3M beams with pegs. Figure 8.65 shows how it should look.

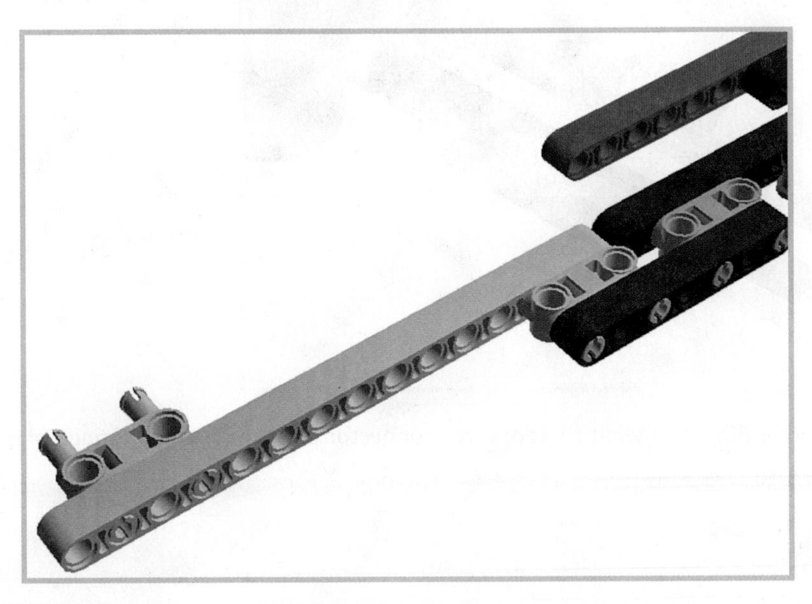

FIGURE 8.65 The course continues to take shape!

64. Add a 5M beam with a couple of connector pegs, as shown in Figure 8.66.

FIGURE 8.66 Attach a 5M beam.

65. Attach another 15M beam to the 5M beam and then throw in a couple extra 3M beam with pegs (see Figure 8.67). Those parts are super handy!

FIGURE 8.67 Another 15M beam gets added to the course.

66. More pegs! Figure 8.68 shows what to add: three connector pegs and a cross connector.

FIGURE 8.68
Add three connector pegs and a cross connector.

67. The course just got a whole lot of support! Add a 13M beam, a 7M beam, and a 5M beam, as shown in Figure 8.69.

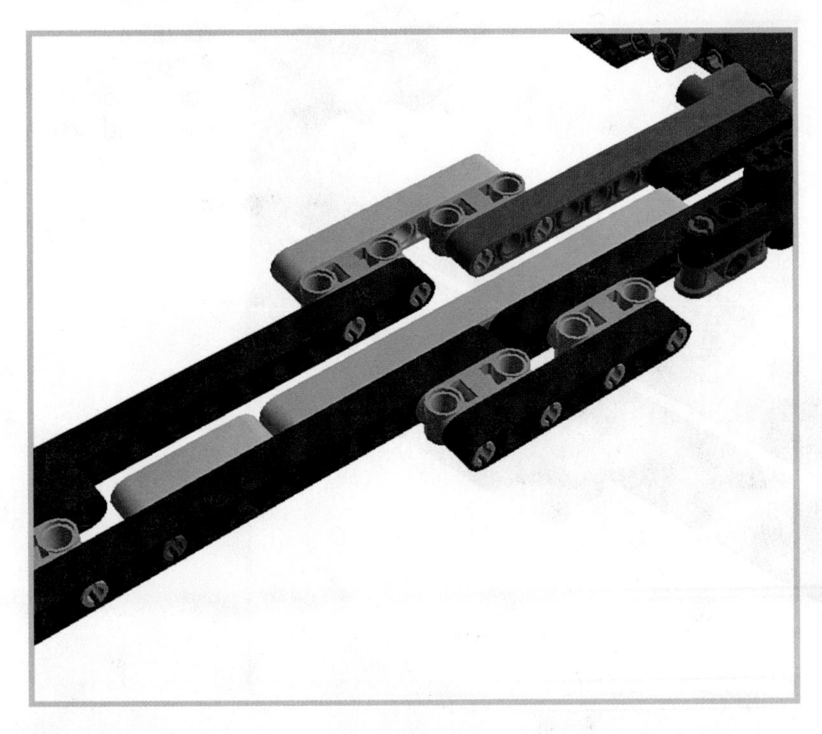

FIGURE 8.69
Add three beams.

68. Add a pair of double cross blocks and secure with cross connectors (see Figure 8.70).

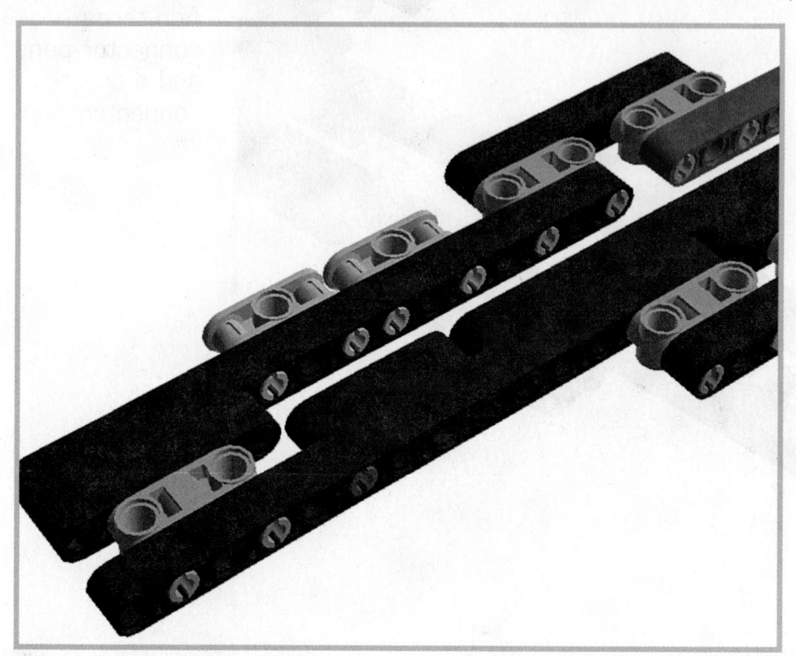

FIGURE 8.70 A pair of double cross blocks get added next.

69. Let's add more pegs! Three each connector pegs and cross connectors get added, as shown in Figure 8.71.

FIGURE 8.71 A bunch of connector pegs and cross connectors get added.

70. Attach two 3x7 angle beams and a double-angle beam, as you see in Figure 8.72. These are the bumpers that help keep the ball on course.

FIGURE 8.72
Add the bumpers.

71. Insert a cross connector and connector peg into the double-angle beam (see Figure 8.73).

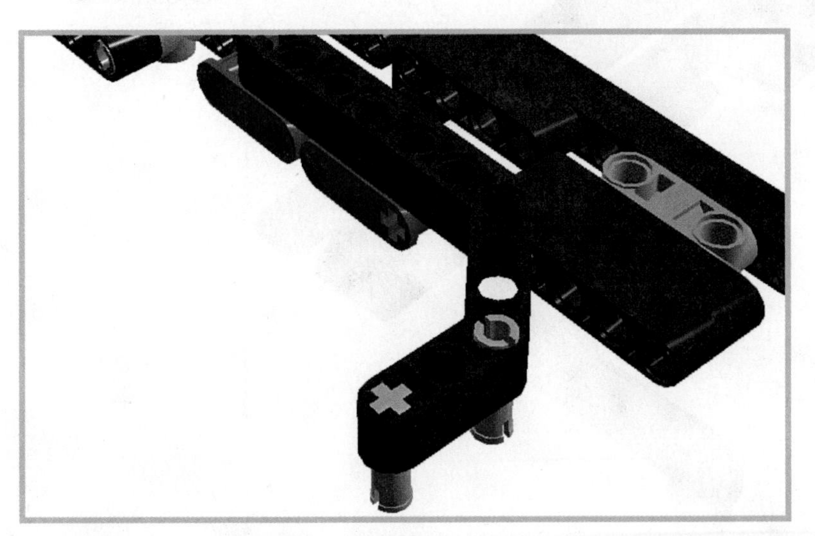

FIGURE 8.73
Add a couple of pegs.

72. Attach a 7M beam to the pegs you just added, as shown in Figure 8.74.

FIGURE 8.74 Attach a 7M beam.

73. Insert a pair of connector pegs as shown in Figure 8.75.

FIGURE 8.75 Insert a pair of connector pegs.

74. Attach a 2x4 angle beam to the exposed pegs (see Figure 8.76). While you're at it, insert a connector peg and a cross connector as well.

FIGURE 8.76
Add a 2x4 angle beam.

75. Attach a 3x5 angle beam to the pegs you just added, and then insert a regular peg and a cross connector. Figure 8.77 shows how it should look.

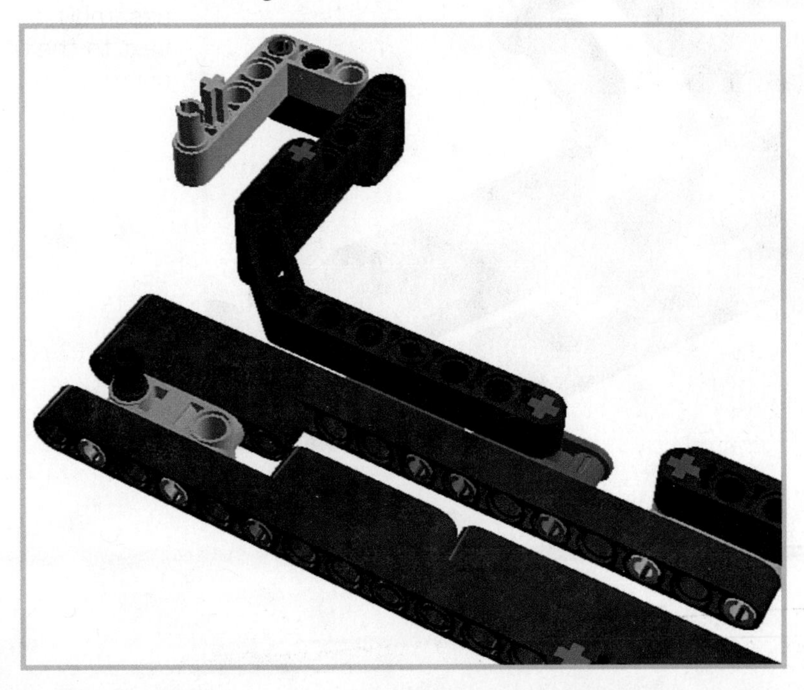

FIGURE 8.77
Add a 3x5 angle beam.

76. Next, let's work on a motor assembly, shown in Figure 8.78. Add four connector pegs and a double-angle beam to the smaller motor.

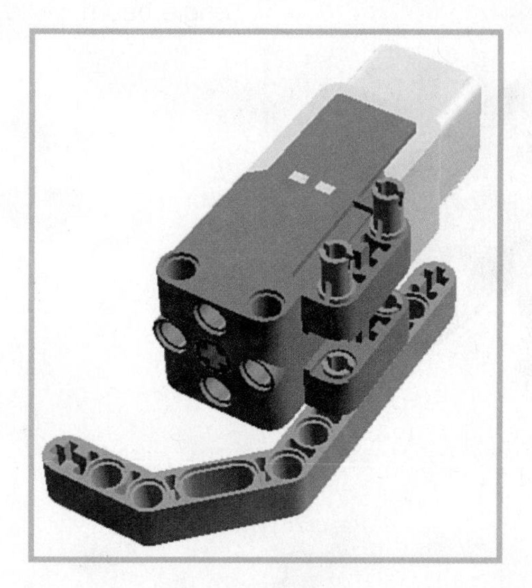

FIGURE 8.78 Add four connector pegs and a double-angle beam to the motor.

77. Place the motor assembly next to the course, as shown in Figure 8.79. It doesn't attach yet!

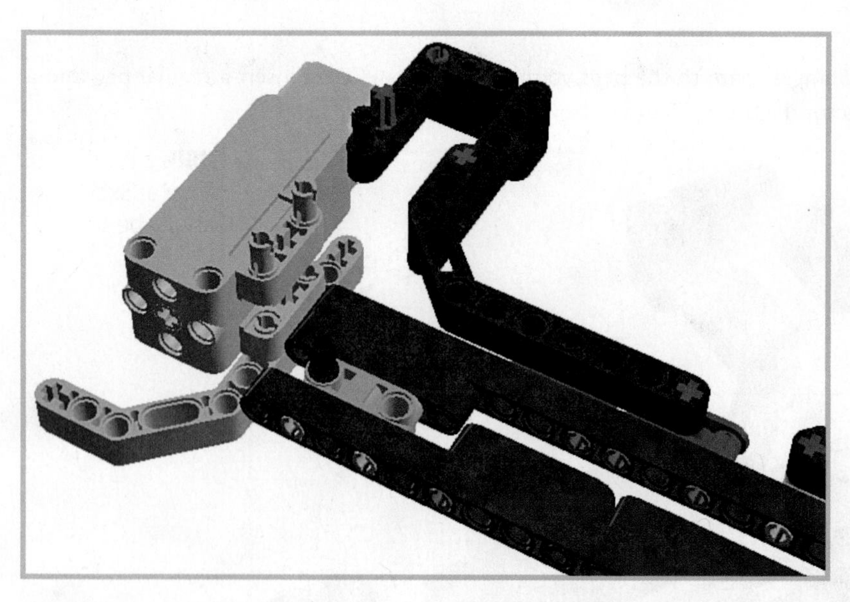

FIGURE 8.79 Set the motor assembly next to the course.

78. Attach a double-angle beam to help support the motor. You can see a top view in Figure 8.80.

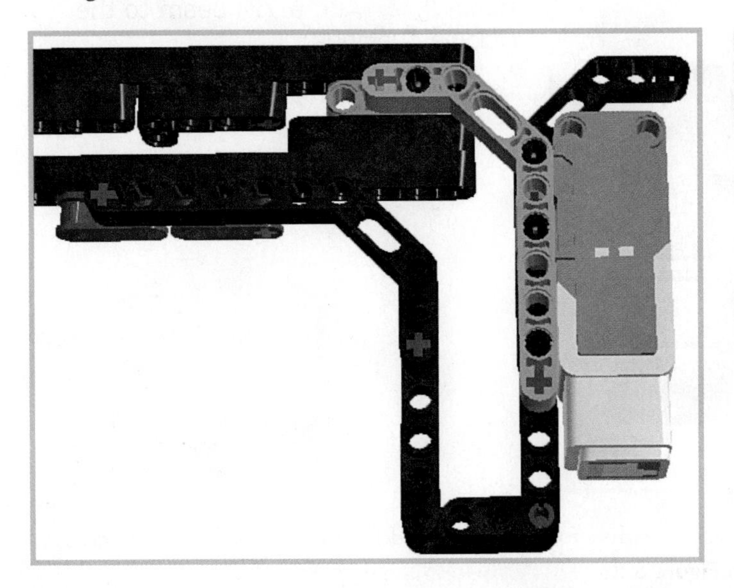

FIGURE 8.80 A double-angle beam helps secure the motor.

79. Attach two module bushings to the double-angle beam, as shown in Figure 8.81.

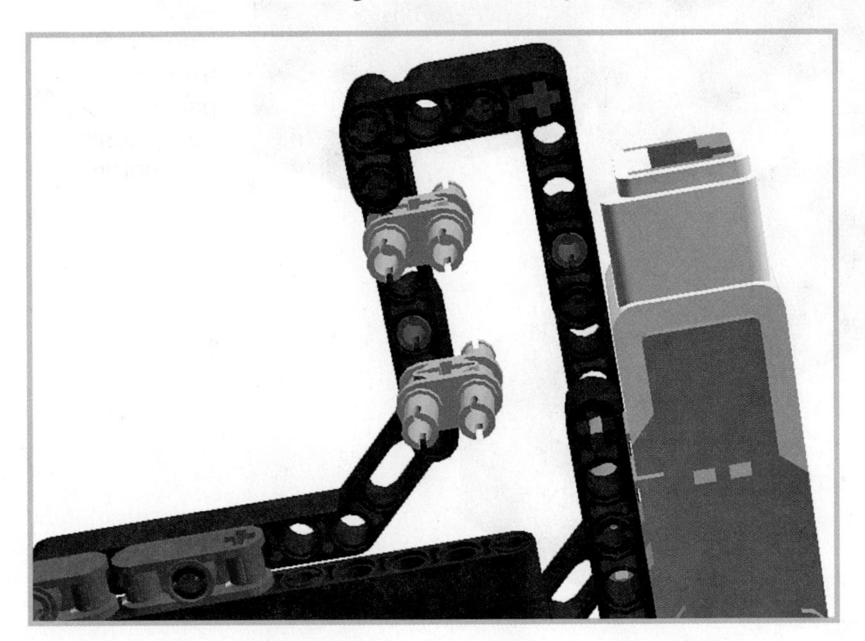

FIGURE 8.81 Add two module bushings.

80. Attach a 7M beam to the free pegs of the module bushings (see Figure 8.82).

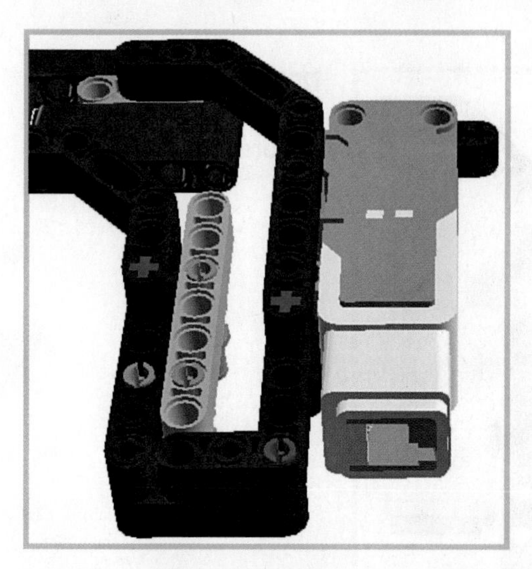

FIGURE 8.82 Add a 7M beam to the module bushings.

81. Next, let's work on the gear assembly for the motor. Add a 3M axle and a pair of cross connectors, as shown in Figure 8.83.

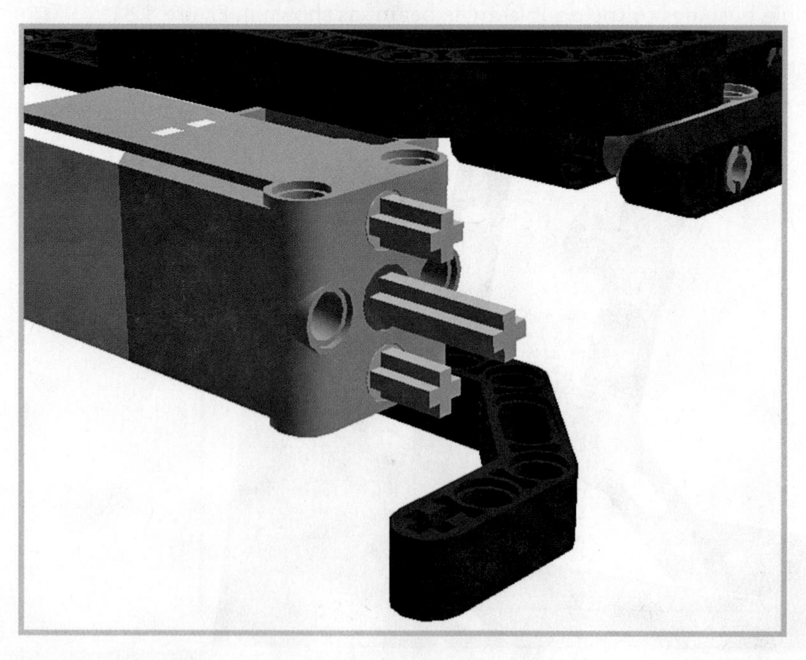

FIGURE 8.83 Insert a 3M axle and a pair of cross connectors to the motor.

82. Attach a 3M beam with fork to the cross connectors, with the axle rotating freely through the center hole of the fork (see Figure 8.84).

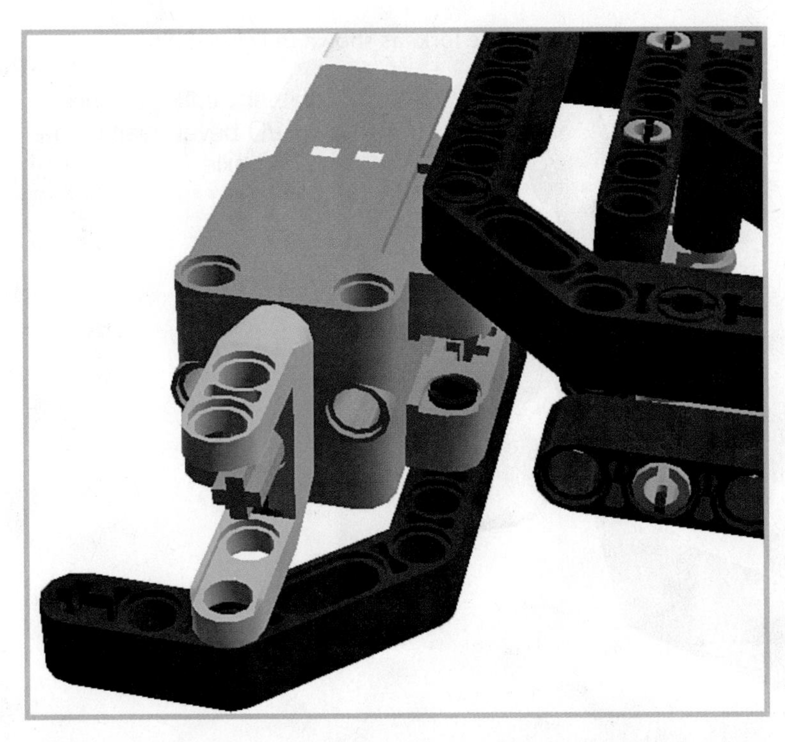

FIGURE 8.84
The 3M beam with fork supports a small gear assembly.

83. Slide a Z12 bevel gear onto the 3M axle. Figure 8.85 shows how it should look.

FIGURE 8.85
Slide a Z12 gear onto the axle.

84. Slide a 5M axle and Z20 bevel gear through the fork, as shown in Figure 8.86.

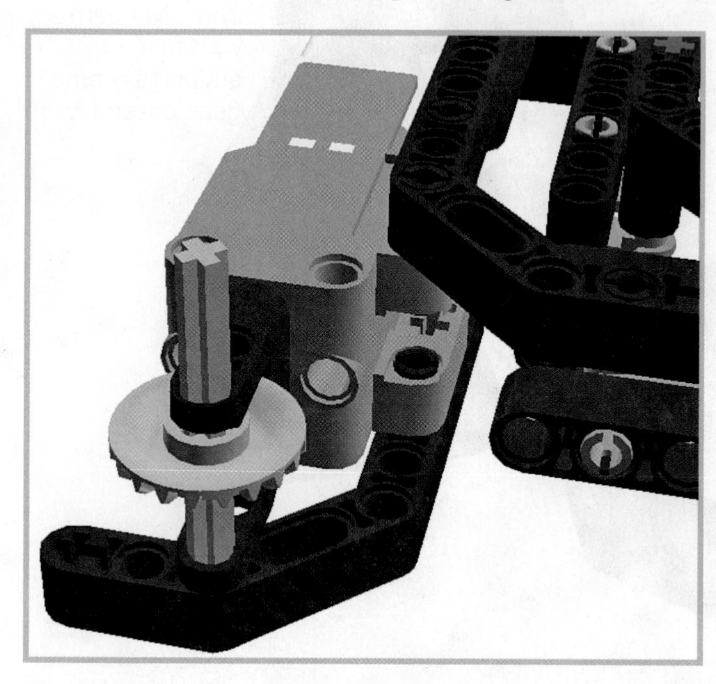

FIGURE 8.86 Secure a Z20 bevel gear with a 5M axle.

85. Slide a Z36 gear onto the 5M axle and then insert a connector peg. Figure 8.87 shows how it should look.

FIGURE 8.87 A Z36 gear slides onto the 5M axle.

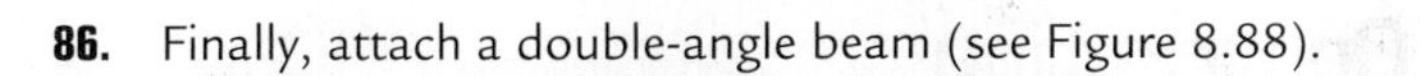

86. Finally, attach a double-angle beam (see Figure 8.88).

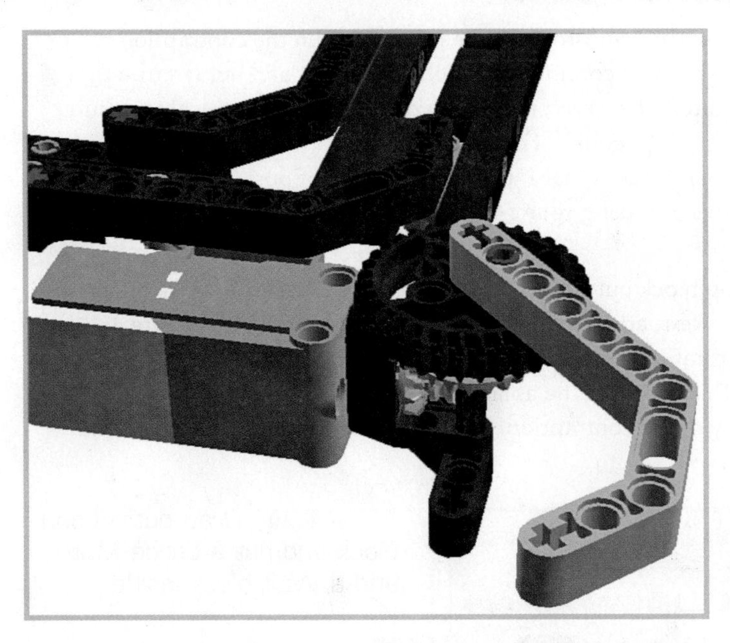

FIGURE 8.88 Add a double-angle beam.

You're done! Figure 8.89 shows the completed contraption with both parts arranged together.

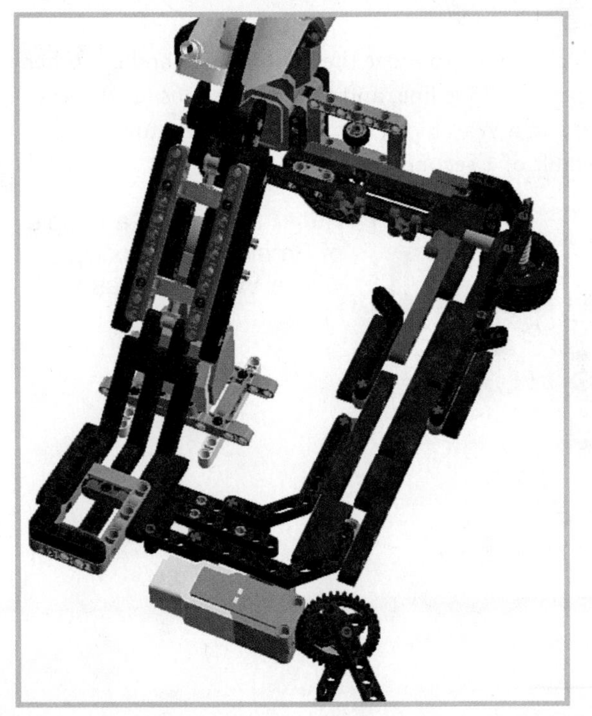

FIGURE 8.89 The contraption is complete!

Program the Contraption

It should not surprise you that the Mindstorms code needed to run the contraption is simple; it's just three motors running continuously, right? Well, yes and no. It's true that there are no sensors, only motors. However, the timing of the motors is critical to ensure that the ball successfully circuits the course. For instance, one of the motors turns an arm that launches the ball down the track. What if the arm were out of position when the ball is delivered to it? Because of this, tweaking your delays and motor speed may prove necessary to get your ball circuiting.

1. Begin by dragging a Loop block out of the Flow palette and place it on the work area as shown in Figure 8.90. Next, add a Large Motor block from the Move palette and set its speed for 30 and its rotations to -1, and set it to port C. Finally, add a Wait block from the Flow palette. This gives you the ability to control the timing of the actions and improve the functionality of the contraption. Leave it at the default of 1 second and adjust it as needed.

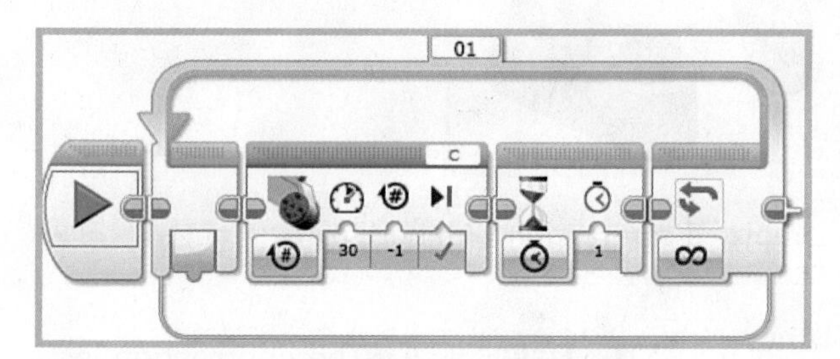

FIGURE 8.90 Drag out a Loop block and put a Large Motor and a Wait block inside.

2. Next, add a couple of Small Motor blocks: One to move the arm one way and the other to move it back again. The default speed of 75 is fine, and set the rotations to -0.5 and 0.5, as shown in Figure 8.91. Finally, add a Wait block to allow you to fine-tune the course's flow, and begin with the default of 1 second.

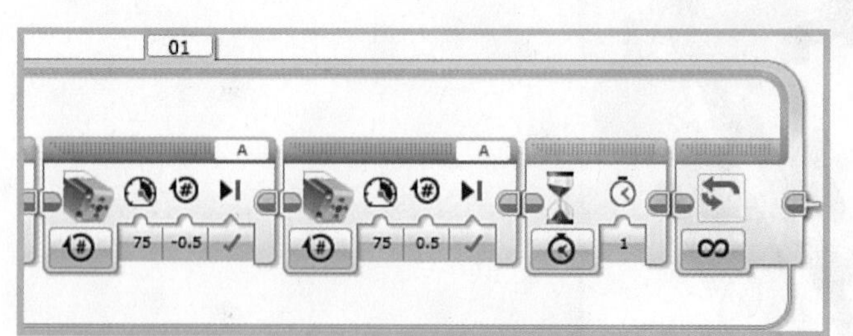

FIGURE 8.91 Add a couple of Small Motor blocks, with a Wait block as well.

3. We work on the teeter-totter in the next two steps. First, add a Large Motor block and set it for port D, 30 speed, and 0.25 rotations. Follow it up with a Wait block just like the others. Figure 8.92 shows how it should look.

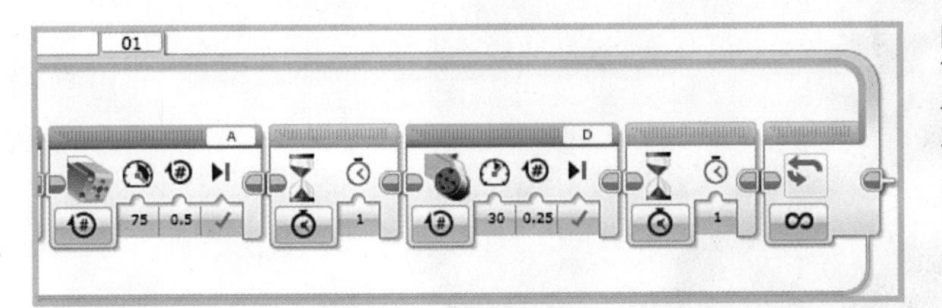

FIGURE 8.92
The teeter-totter is rotated forward in this step.

4. Finally, we need to return the teeter-totter to its default position. Add another Large Motor block and change the rotations to -0.25, following it up with one last Wait block (see Figure 8.93).

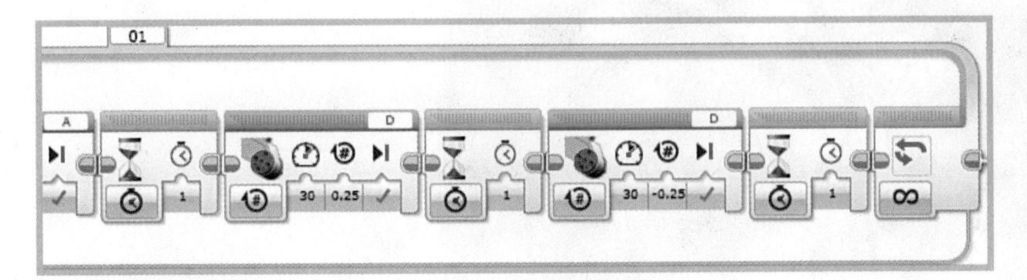

FIGURE 8.93
The teeter-totter returns to its default position.

You're done! Run the program and begin learning how to tweak the motor speeds and delays to keep the ball in motion.

Creating Your Own Parts

With a complicated model like this, a builder can identify a number of possible areas in which a custom part would be useful. However, I focused on just two: I wanted to manufacture the perfect "gear" that would move the ball around the track, and I also wanted a baseplate onto which I could mount the robot, as its current structure is a bit wobbly. I simply ran out of beams to reinforce the robot, limited as I was by using only one set to build it. But that's where designing and outputting your own parts becomes even more useful! You can see both parts I created in Figure 8.94.

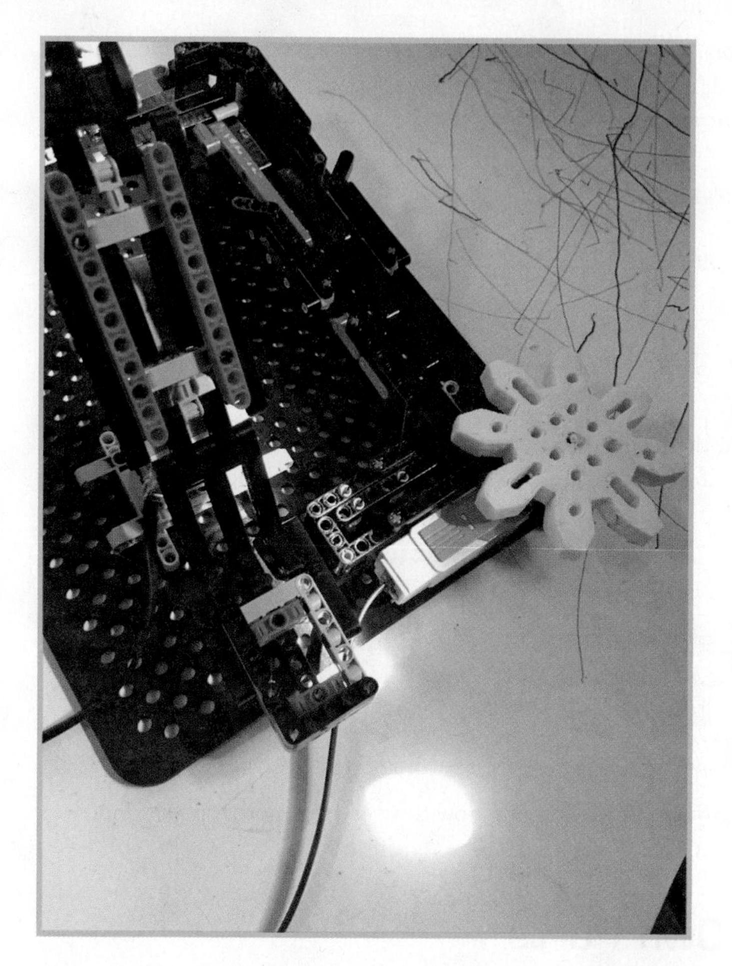

FIGURE 8.94 The contraption, augmented with additional parts.

The Baseplate

I'm going to work on the baseplate first. It's simply a wooden board pierced with Mindstorms-compatible mounting holes, allowing us to attach the contraption to it for added security (see Figure 8.95). This is a great solution if you have a robot you want to protect from breaking, because a secure base makes everything attached to it more secure! Let's face it, plastic beams secured with plastic pegs will *always* have some give—sometimes too much! Finally, this plate will come in handy for years to come because you can use it for other projects.

FIGURE 8.95 The baseplate is a great platform upon which to build a robot.

Parts

The parts needed for the project are modest, not counting the laser cutter!

- **Wood**—I use 1/4-inch Baltic birch plywood when lasering.
- **Mounting hardware**—I used #8-32 socket-headed screws with washers and nuts.
- **Spray paint**—Use your favorite color! I used Krylon chalkboard paint because I love that slate texture, but use what you want.

Steps

To create your baseplate, follow these steps:

1. Measure and determine size and shape. You might want to bust out the ruler and simply measure what you think will be the right size, but in the LEGO world we measure in mounting holes. As mentioned in the previous chapter, the mounting holes used in Mindstorms and Technic sets are 6mm in diameter and spaced 8mm apart. For my baseplate I elected to use only every other hole, so as not to make the board too fragile.

   When determining the outer dimensions of the baseplate, simply figure out the number of mounting holes you want and then add a half-inch or so onto each side. I came up with a 10x13-inch plate with 18 rows of 19 holes, as shown in Figure 8.96.

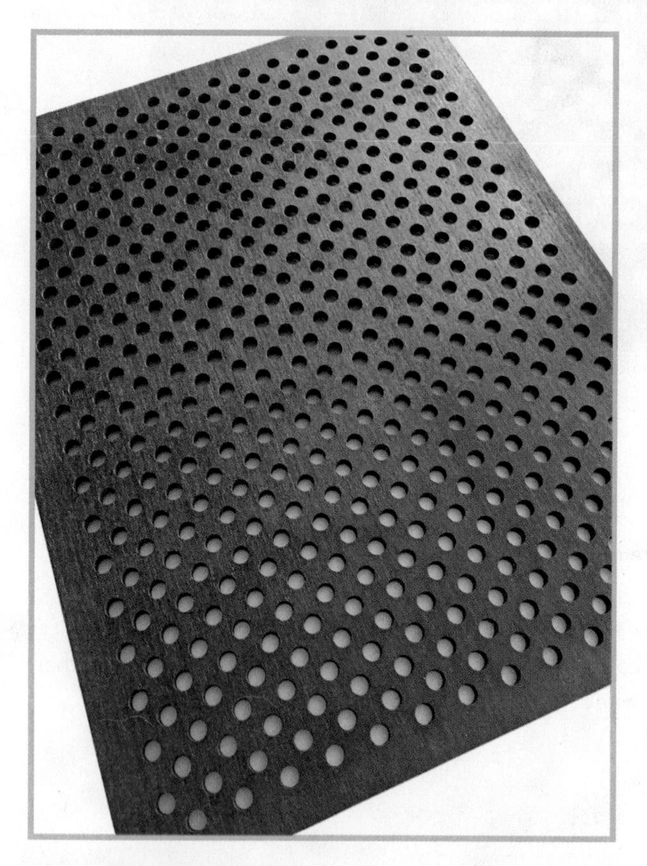

FIGURE 8.96 Determine the dimensions of your plate.

2. Go into Inkscape and create a new document (see Figure 8.97). Go into File > Document Properties and change the canvas size to 10x13 inches, or however large you want your board to be.

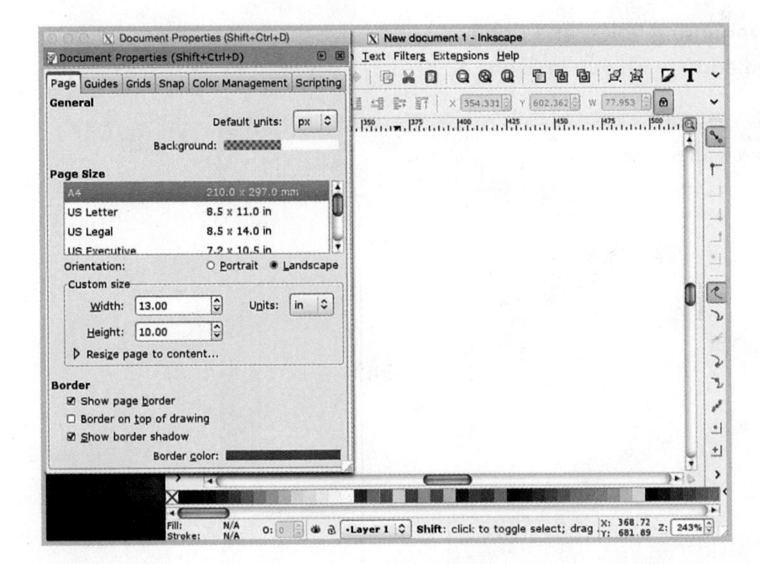

FIGURE 8.97 Create a new document in Inkscape.

3. Use the circle tool to make a 6mm circle, shown in Figure 8.98. This is your first mounting hole! Use the X and Y coordinates to give it a logical spot on your document. I randomly chose the coordinates of 100 by 170, and placed the first circle there. The next circle gets placed at the X of 108 (remember, the centers of two adjacent holes are 8mm apart) but keeps the Y of 170.

   Keep doing this! You should be able to pepper your board with as many mounting holes as you want. When you fill up a row, select the whole bunch, hit Edit > Duplicate to duplicate it, and then change the Y to 178. *Voila*, two rows!

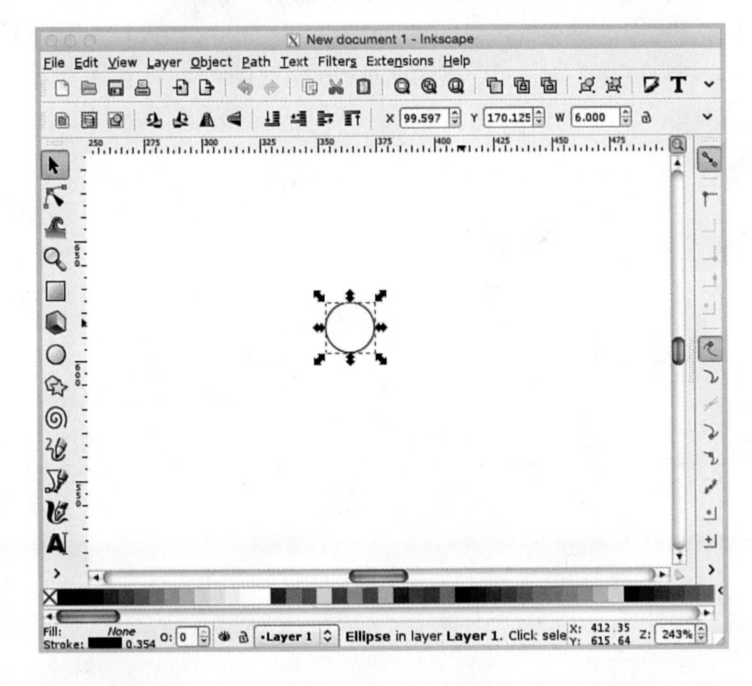

FIGURE 8.98 Draw a circle using Inkscape's circle tool.

4. Cover the board with mounting holes, and don't forget to give the board an edge. I gave mine pleasant rounded corners, shown in Figure 8.99.

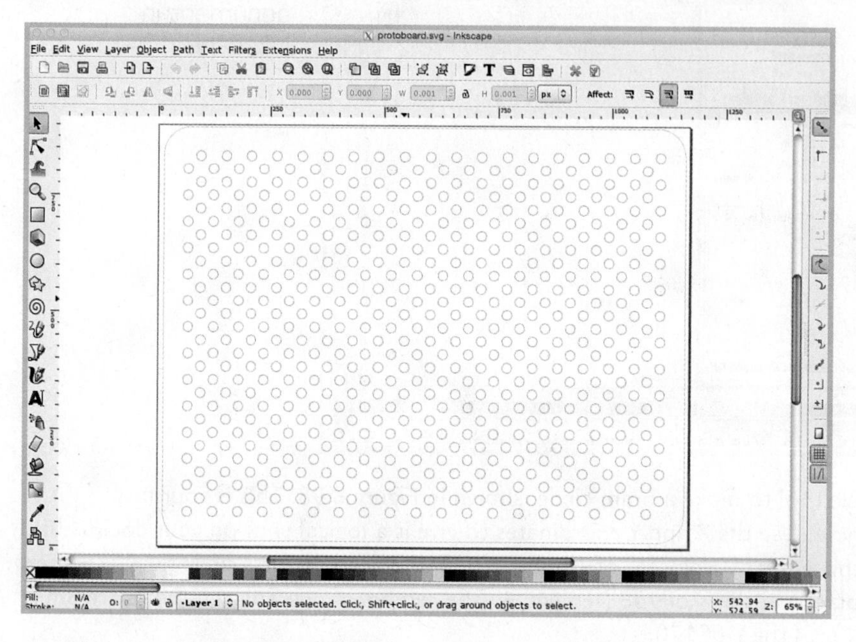

FIGURE 8.99 The design is ready to cut.

5. The .SVG (scaled vector graphics) file Inkscape saves works well with my laser setup, but you may have to convert the file to another format. Consult with your friendly local laser admin! Figure 8.100 shows the Save screen.
6. Once your design is complete, cut out the plate on your laser cutter (see Figure 8.101).

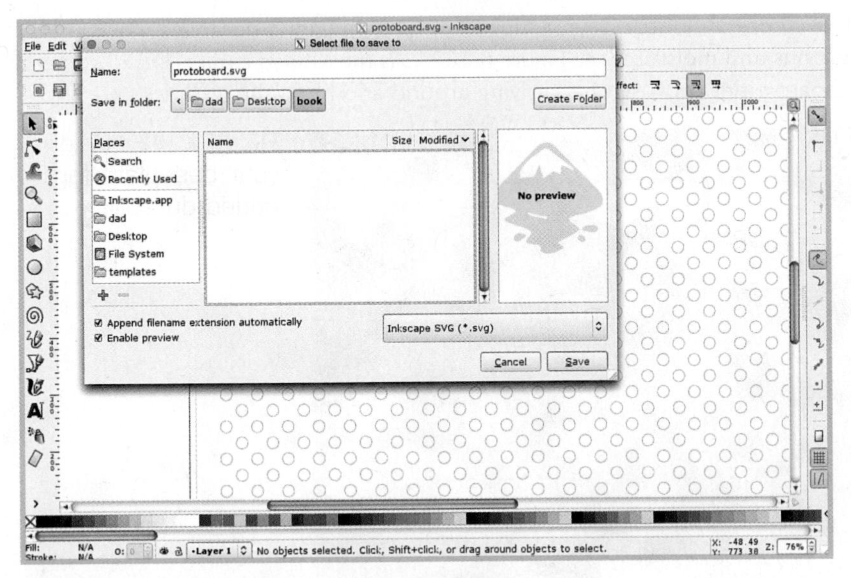

FIGURE 8.100 Save the design as a .SVG.

FIGURE 8.101 The laser cuts out the baseplate.

7. Once the plate comes out of the laser, I suggest painting it with a durable spray paint to help it resist stains and moisture (see Figure 8.102). As mentioned earlier, I used some Krylon chalkboard paint because I had it lying around and I love the texture.

FIGURE 8.102 Paint your baseplate for added durability.

8. Install your Mindstorms components to the plate (Figure 8.103) using LEGO parts (such as axles and bushings) or the #8-32 screws I mentioned.

FIGURE 8.103 Attach the contraption to the baseplate.

The Gear

The other custom part I'm going to build for this project is a plastic gear that helps roll the ball through the course. The gear consists of a knobbed shape with Mindstorms mounting holes and a cross-axle hole piercing it. You can see it in Figure 8.104.

One question you might be asking is, why 3D print? I suppose you could laser the shape out of quarter-inch plywood or something similar, but for this exercise I'm showing you how to turn it into a 3D-printed design because the next part you output might be not at all laserable.

FIGURE 8.104
The 3D-printed gear helps keep the ball moving along the course.

Parts

Other than a 3D printer and filament for it, you won't need anything to create the gear!

Steps

To design and output your own gear, follow these steps:

1. Determine the gear shape and size. I went with an eight-toothed, knobby gear that was 4 inches in diameter, as shown in Figure 8.105, but make yours look the way you want it! The main consideration is whether it helps move the ball along the course.

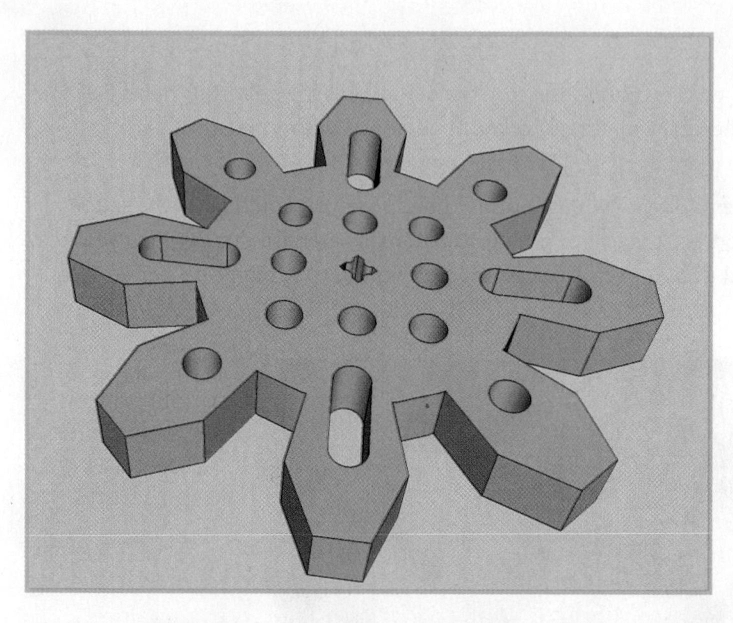

FIGURE 8.105 Decide on the size and shape of the part.

2. Design it in Inkscape (see Figure 8.106). Create the part the same way as the baseplate. However, we'll be making it three-dimensional in a different program. For best effect, make the center hole a cross-hole.

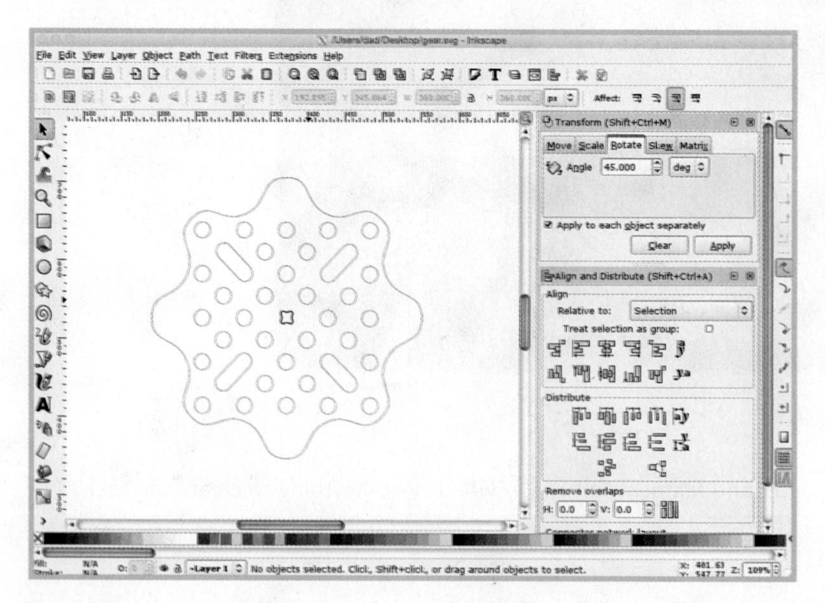

FIGURE 8.106 Design the gear in Inkscape just as you did the baseplate.

3. Bring the design into SketchUp. The free version of SketchUp doesn't permit the import of vector art. It's a feature that has been intentionally crippled to get people to pay for the Pro version of the program, which does allow it. If you just have the free version, import a PDF version of the Inkscape design (see Figure 8.107). The PDF appears as a

raster graphic—as pixels rather than lines—so you have to trace over each item. It's there just to give you an idea of where to place things.

Another solution involves importing the Inkscape file into Tinkercad, a 3D-design application. From there you can use its Shape Generator function to make the object 3D.

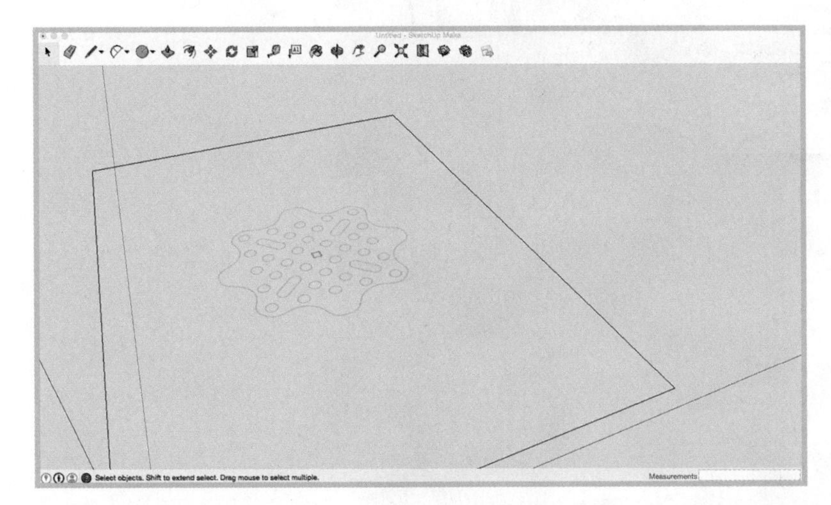

FIGURE 8.107 Import the PDF into SketchUp.

4. Once you have re-created the design in SketchUp, use the Push/Pull tool to "grab" the top and pull it up, as shown in Figure 8.108. You can use your ruler tool to measure the width. I went with around 10mm.

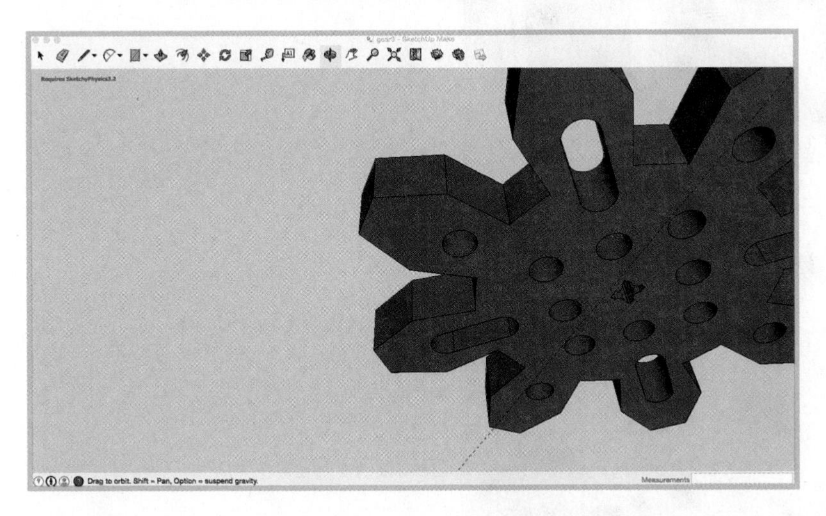

FIGURE 8.108 The push/pull tool makes things 3D.

5. Export the gear. You want to export in a format your 3D printer supports, but I ended up going with the .STL format (which means "stereolithography") that my 3D printer's software likes. Figure 8.109 shows the export in progress.
6. Print! Figure 8.110 shows me printing the part on my 3D printer.

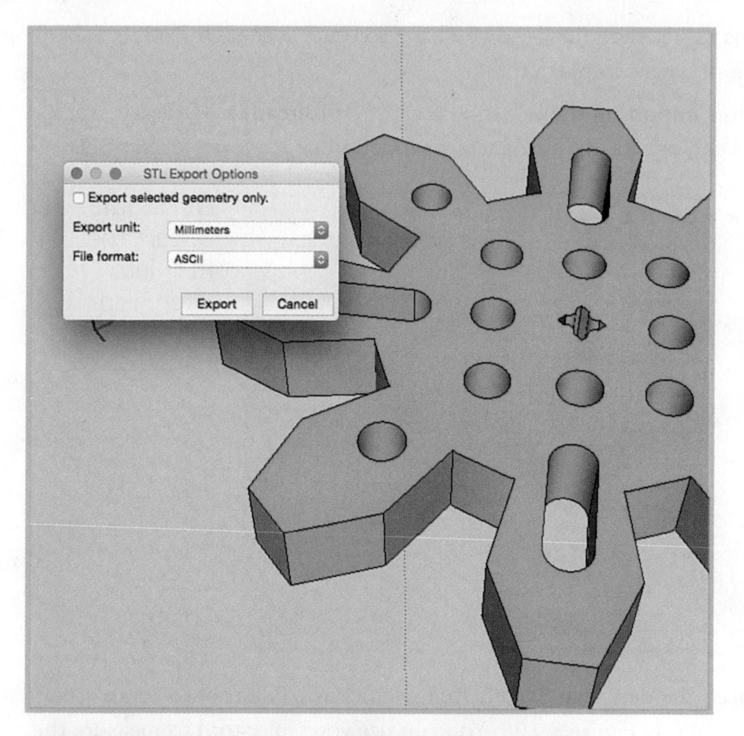

FIGURE 8.109 Export the gear as a .STL.

FIGURE 8.110 Print off the part!

7. Install the gear. If you designed the cross-hole correctly, it should slide on just like any other gear, as you can see in Figure 8.111.

FIGURE 8.111 Install the gear onto the contraption.

Summary

I love how this chapter actually gave you a practical reason to create new parts and then walked you through doing it! In Chapter 9, "Hacking LEGO IV: Add-on Electronics," you learn how to further modify your EV3 set with the addition of third-party sensors, lights, and other electronic gadgetry.

Hacking LEGO IV: Add-on Electronics

I already talked about how people modify Mindstorms to suit their own needs. Some of the more successful of these projects become products so other folks can play with the same ideas, only without the time and money invested in development.

Just to be clear, in other chapters I mention how microcontrollers such as the Arduino can be utilized to control Mindstorms models, and these devices can natively control a plethora of sensors and motors. However, in this chapter I mostly talk about add-on electronics that can be plugged directly into the EV3 Intelligent Brick and used without too much modification; nothing more major than installing some new software on the brick.

Figure 9.1 shows an ad-hoc modification that well-known LEGO expert Joe Meno made to his EV3 brick: He added a light strip to see the display, because it's not backlit and sometimes is hard to read. This is a classic example of hacking the EV3 to improve it, encountering and correcting an area where the product doesn't work well for you.

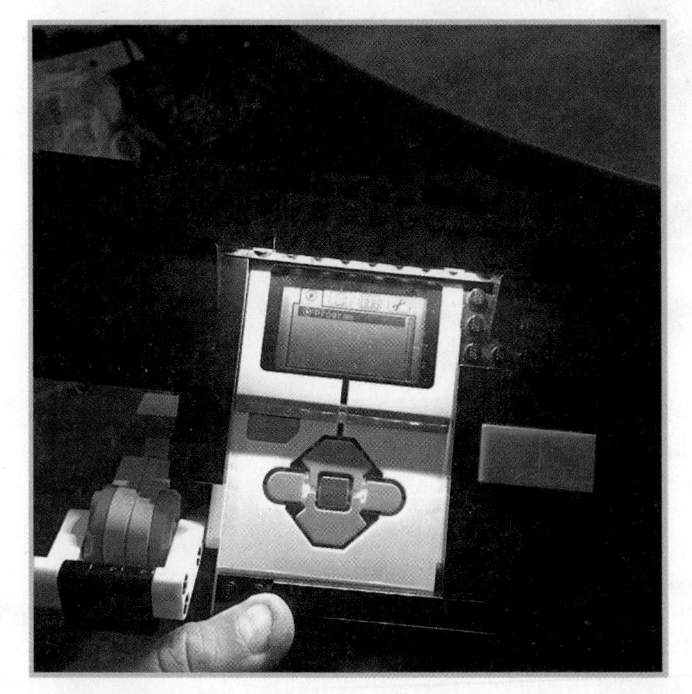

FIGURE 9.1 Need a backlight? Just add your own. Credit: Joe Meno

Motors and Wheels

While I'm mostly visualizing sensors and similar components, motors and wheels may also be purchased to make your robot even cooler. Additionally, you often need some sort of board to control motors, and in the following section you find an example.

Motor Driver

Wayne and Layne created Bricktronics to interface Mindstorms components with the Arduino world. The Motor Driver represents their latest effort and ditches the Arduino altogether—at least on the interface board. The platform-neutral pinouts allow you to swap in a Raspberry Pi or BeagleBone Black, or just use an Arduino to control a robot's motors. Figure 9.2 shows the Motor Driver in use.

https://www.wayneandlayne.com/projects/bricktronics-motor-driver/

FIGURE 9.2 Wayne and Layne's Motor Driver allows you to control Mindstorms motors with an Arduino or Raspberry Pi.

Omni-Wheels

Okay, these Rotacaster omni-wheels are not electronic, but they are extremely cool-looking as well as Mindstorms-compatible (see Figure 9.3). Omni-wheels have mini wheels along the rim of a big wheel, allowing the robot to roll perpendicularly to the main axis of the robot. A common application of this involves placing two sets of wheels 90 degrees from each other, and one set drives the robot while the other set sits idle with its unpowered mini wheels turning freely.

http://www.rotacaster.com.au/

FIGURE 9.3 Omni-wheels allow for lateral movement as well as forward and backward. Credit: Rotacaster

8-Channel Servo Controller

This small board mates a standard Mindstorms plug with eight 3-prong servo motor plugs, allowing the Intelligent Brick to easily control these small motors (see Figure 9.4). Mindsensors includes control blocks for both EV3 and NXT, as well as sample programs for a variety of platforms including RobotC and LabView.

http://www.mindsensors.com/index.php?module=pagemaster&PAGE_user_op=view_page&PAGE_id=93

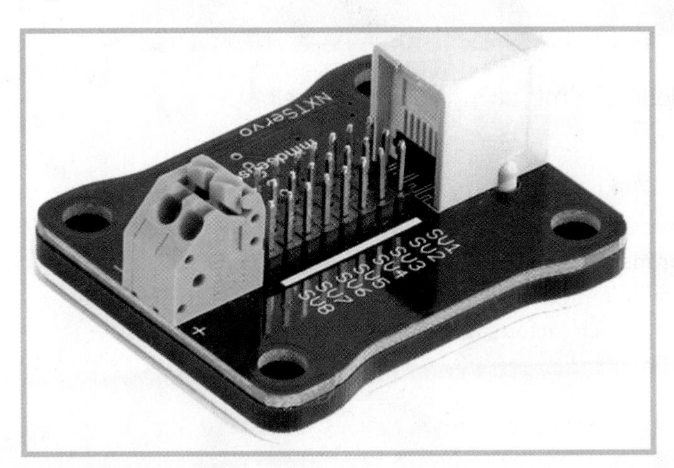

FIGURE 9.4 This small board allows the EV3 Intelligent Brick to control up to eight servo motors. Credit: Mindsensors

Hub-ee Wheels

Creative Robotics' Hub-ee wheels are so-called "hub wheels" that contain a servo and gearbox inside the wheel itself, saving space on the main robot (see Figure 9.5). The wheels have Mindstorms mounting holes, but the EV3 brick cannot natively control the servos. Of course, a variety of products allow this—such as the 8-Channel Servo Controller I just mentioned!

http://www.creative-robotics.com/

FIGURE 9.5 Hub wheels enclose the motor in the wheel itself.

Linear Actuator

A linear actuator moves a rod back and forth, rather than rotating it. They tend to be slow and strong, as a worm gear inside the actuator does the work. LEGO sells its own actuator (P/N 5003110, shown in Figure 9.6), but there are others, such as Firgelli Industries' (firgelli.com) offerings, which include metal-sheathed and Mindstorms-compatible actuators.

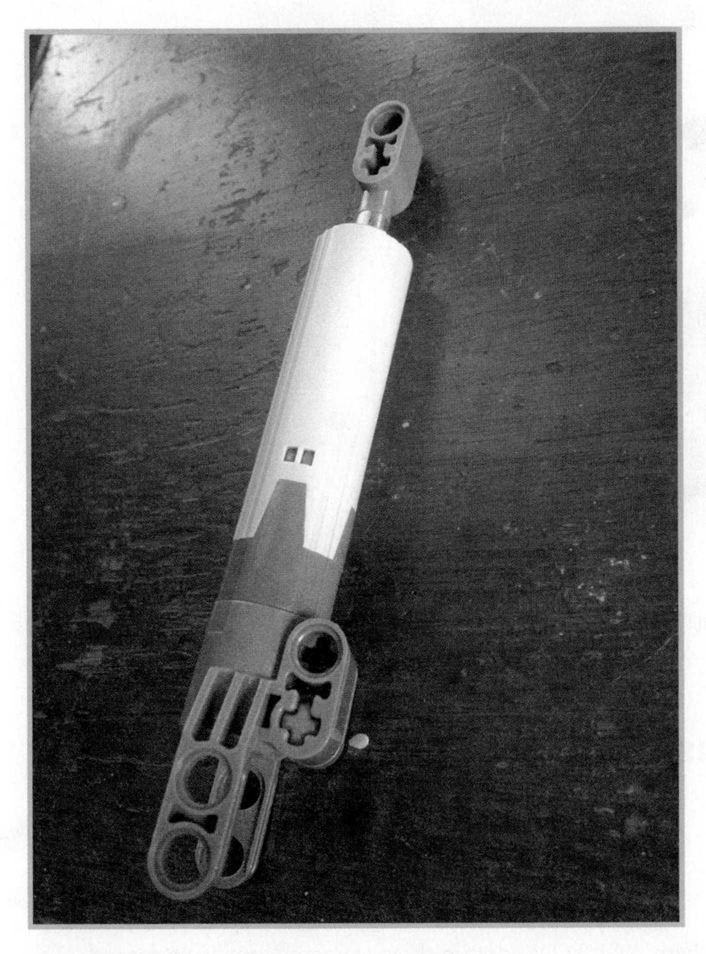

FIGURE 9.6 A linear actuator extends and retracts a rod rather than rotating it.

Servo-Actuated Pneumatic Valve

Mindsensors offers a pneumatic valve that can be used to interface the EV3 Intelligent Brick with LEGO's pneumatics set (see Figure 9.7). Combined with a pressure sensor, the EV3 can fully control the set's system of pressure tanks and hoses.

http://www.mindsensors.com/index.php?module=pagemaster&PAGE_user_op=view_page&PAGE_id=141

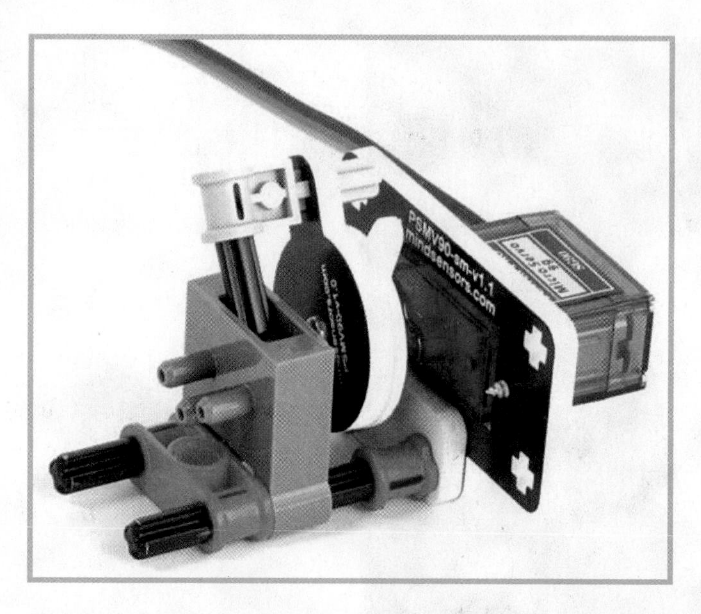

FIGURE 9.7 This servo-controlled valve can be triggered by the EV3 brick.
Credit: Mindsensors

Lighting Systems

These components add a little illumination to your next project. (Often they're just there to make the robot look good.)

RGB LED Modules

Some LEDs are cooler than others. RGB LEDs contain three elements, one each for red, green, and blue, and the three colors can be combined to create any hue. Dexter Industries dLights are chainable lights, with a base module that plugs into the EV3 brick (see Figure 9.8). Up to three satellite LED modules can be added on to the base.

http://www.dexterindustries.com/shop/dlight-led-lights-lego-mindstorms-nxt-ev3/

FIGURE 9.8 These RGB LED modules are controlled by the EV3 brick. Credit: Dexter Industries.

Light Strands

Brickstuff sells an Arduino-controlled lighting system for LEGO robots, featuring ultra-thin wires and tiny LEDs that fit inside many elements. While designed for LEGO's classic System bricks (the ones with the studs, shown in Figure 9.9), you can also incorporate Brickstuff into a Mindstorms project.

https://www.brickstuff.com/

FIGURE 9.9 Add tiny lights to your next robot. Credit: Brickstuff.

Sensors

Unsurprisingly, many of the add-on modules available for sale consist of sensors not found in LEGO Mindstorms' normal catalog.

Grove Sensor Adapter

The Grove System consists of a series of sensors built into PCBs that use Molex plugs to connect to other parts of the circuit (see Figure 9.10). Mindsensors' Grove Sensor Adapter plugs into an EV3 and allows it to take readings from these components.

http://www.mindsensors.com/index.php?module=pagemaster&PAGE_user_op=view_page&PAGE_id=209

FIGURE 9.10 The Grove system of sensors can be harnessed by the EV3 brick, thanks to this adapter. Credit: Mindsensors.

dGPS

Dexter Industries' dGPS sensor receives GPS coordinates and sends latitude, longitude, time, speed, and heading to the EV3 or Raspberry Pi (see Figure 9.11). In addition to being used natively by the EV3, and programmable via Python, C, or Java, you can also control the dGPS in EV3 bricks that have been refreshed with alternate firmware packages RobotC and Lejos.

http://www.dexterindustries.com/shop/dgps/

FIGURE 9.11 The dGPS gives your robot the capability to receive GPS information. Credit: Dexter Industries.

Proximity Sensor

This proximity sensor is designed to work with LEGO trains, sending a signal when the train passes overhead (see Figure 9.12). It could also be used for other proximity sensing needs, though it's not natively compatible with EV3. This product is in development but will be available at brickstuff.com at some point.

https://www.brickstuff.com/

FIGURE 9.12 Brickstuff's proximity sensor detects when LEGO trains pass overhead. Credit: Brickstuff.

Pixy and Pixy Adapter

The Pixy is an inexpensive camera module that connects to a Raspberry Pi or Arduino. It includes a cradle for the camera's PCB and a Mindstorms plug so you can plug it directly into your EV3 brick (see Figure 9.13).

http://www.mindsensors.com/index.php?module=pagemaster&PAGE_user_op=view_page&PAGE_id=215

FIGURE 9.13 The Pixy camera is designed to be added onto robots outfitted with the Pixy Adapter. Credit: Mindsensors.

Sensor Mux

Mindsensors offers a *mux*, which is a short way of saying *multiplexer*. This is a device that allows multiple sensors to send information to a single EV3 port, allowing an Intelligent Brick to control far more than the normal four sensors (see Figure 9.14).

http://www.mindsensors.com/index.php?module=pagemaster&PAGE_user_op=view_page&PAGE_id=134

FIGURE 9.14 The Sensor Mux allows you to take in sensor data from multiple sources while only using one port. Credit: Mindsensors.

Absolute IMU-ACG

Mindsensors' accelerometer measures tilt and compass direction to 1 degree, as well as acceleration to 10 millionths of a G (see Figure 9.15). This is just the ticket to make a self-balancing robot, for instance, which rights itself when it detects that it's falling over.

http://www.mindsensors.com/index.php?module=pagemaster&PAGE_user_op=view_page&PAGE_id=169

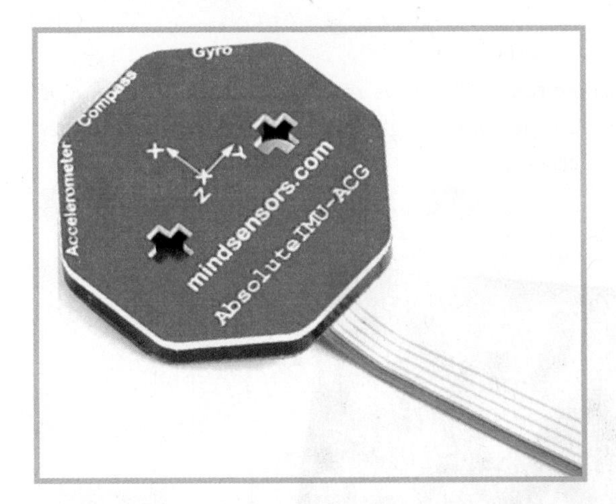

FIGURE 9.15 This sensor detects tilt, compass direction, and acceleration. Credit: Mindsensors.

dPressure Sensor

Dexter Industries' dPressure sensor measures pneumatic pressure. LEGO Education offers a pneumatics add-on set, though it doesn't interface with the EV3 brick. Thanks to Dexter, now it does! You can see the sensor in Figure 9.16.

http://www.dexterindustries.com/manual/dpressure/

FIGURE 9.16 The dPressure sensor tells the EV3 brick the status of a pneumatic system. Credit: Dexter Industries.

Thermal Infrared Sensor

This noncontact heat sensor reads the surface temperature of an object without touching it (see Figure 9.17). This is useful in cases where the robot might be damaged by whatever it's measuring. It can detect a flame within 2 meters.

http://www.dexterindustries.com/shop/thermal-infrared-sensor-lego-mindstorms-nxt-and-ev3/

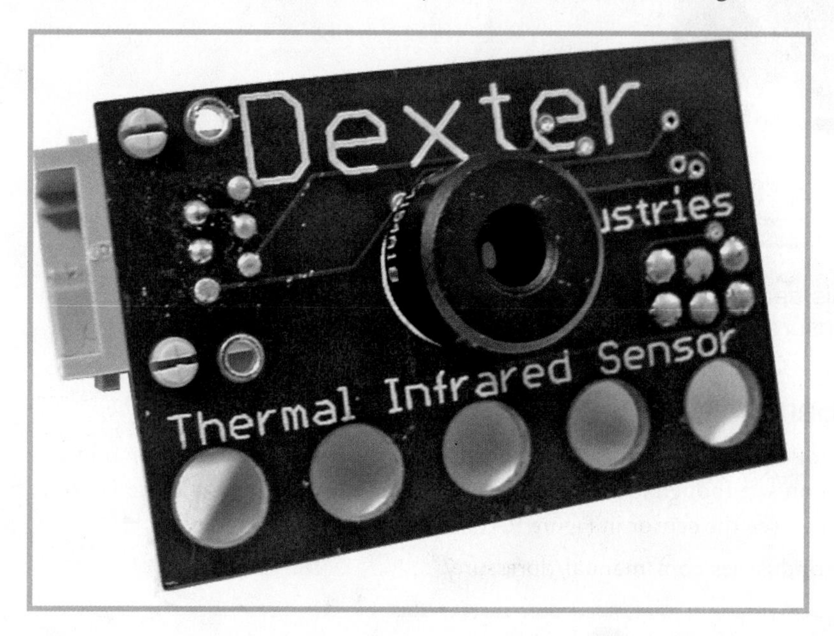

FIGURE 9.17 The thermal infrared sensor can tell how hot something is without touching it. Credit: Dexter Industries.

NXT and Education Electronics

The NXT is the previous version of the EV3, as I've likely mentioned before. Some legacy NXT sensors and motors are compatible with the EV3, and I list a few in this section. Another source of Mindstorms-compatible parts that don't come in the main set is LEGO Education, a site geared toward robotics and science teachers. In the following sections I include some products available for purchase from that site as well.

Pneumatics Add-On Set

LEGO hackers love its pneumatics set, which includes an air tank, pumps, and tubes. Other products include Mindstorms-compatible pneumatic switches and actuators that work on air pressure. LEGO doesn't support the set as enthusiastically as those hackers would like, but they are a small market! You can see the set in Figure 9.18.

https://shop.education.lego.com/legoed/en-US/catalog/ product.jsp?productId=9641

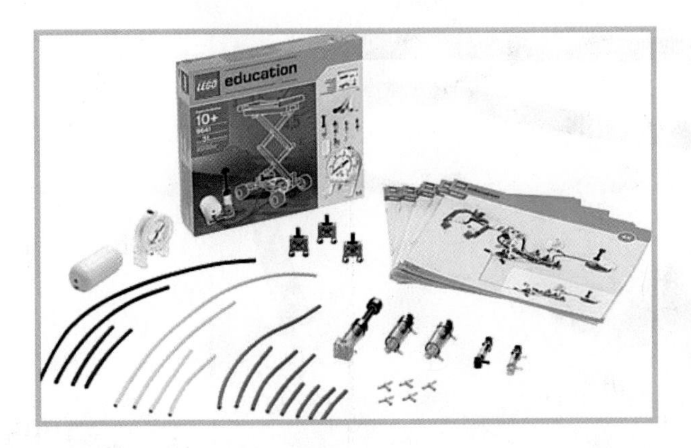

FIGURE 9.18 Add pneumatics to your next project.

NXT Light Sensor

This sensor detects the intensity of light and returns a value to the Intelligent Brick (see Figure 9.19). The EV3 set has a robust light sensor that also can detect the color of the item scanned. Why would someone want to use this old sensor? Not sure—maybe they have one lying around and want to add it to their next project.

FIGURE 9.19 The NXT's light sensor is old school, but might come in handy.

NXT Temperature Sensor

LEGO Education rightly plays up Mindstorms' science applications and supports it with sensors such as the one shown in Figure 9.20. The temperature measured can be expressed as centigrade or Fahrenheit, and reads between -4° F/-20° C and 248° F or 120° C.

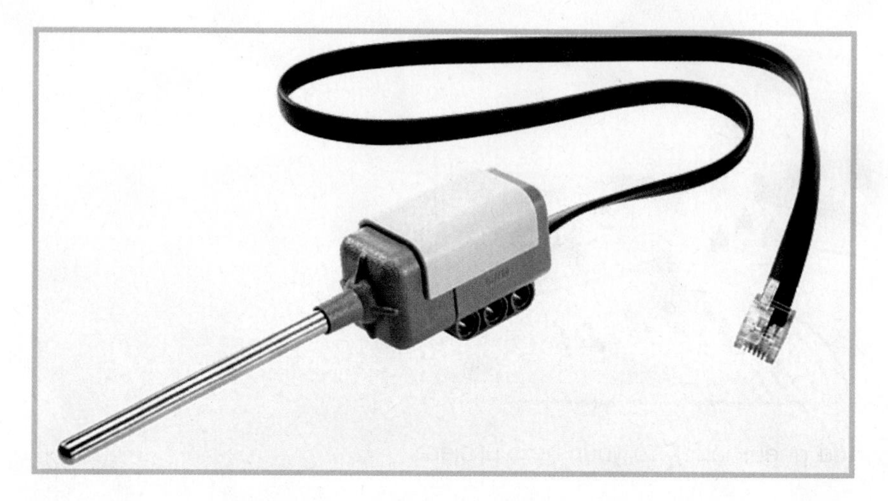

FIGURE 9.20 Measure temperature with this sensor. Credit: LEGO Education.

NXT Sound Sensor

Like the light sensor mentioned previously, the sound sensor is another orphaned NXT sensor (see Figure 9.21). Unlike the light sensor, however, LEGO hasn't seen fit to replace it with something better. As such, you'll find yourself reaching for one of these if you need to have your robot detect the intensity of sounds.

FIGURE 9.21 The sound sensor detects the volume of nearby sounds.

NXT Ultrasonic

An ultrasonic sensor uses human-inaudible sound waves to measure distances. It sends out a ping and measures the microseconds it takes the ping to bounce back, using simple math to determine the distance (see Figure 9.22). This is another sensor that got put out to pasture with the main EV3 set. LEGO still manufactures them, but they are only sold individually and through the Education version of the EV3 set.

FIGURE 9.22 Use an ultrasonic sensor to measure distances.

Control Systems

How will you control your robot? This section has a few products that may prove useful in this regard.

Relay Driver

Mindsensors sells an EV3-controllable relay, which is kind of an electronic switch for triggering electrical loads far stronger than a low-power microcontroller can handle. For instance, a relay could trigger an assembly of dozens of LEDs, keeping the high voltage power supply away from the Arduino or EV3, which are both rated for a maximum of 5V. You can see the relay in Figure 9.23.

http://www.mindsensors.com/index.php?module=pagemaster&PAGE_user_op=view_page&PAGE_id=140

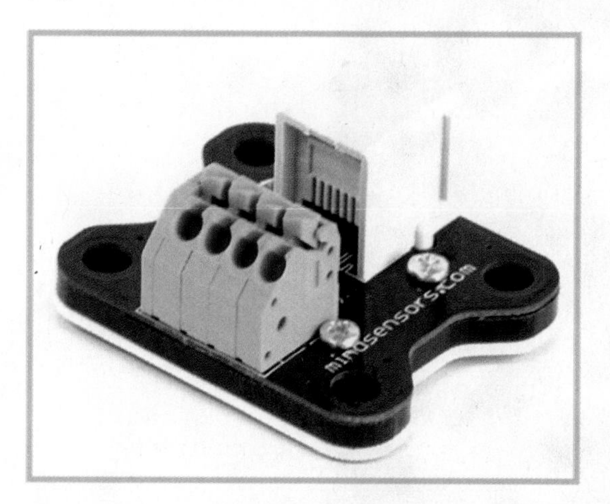

FIGURE 9.23 Switch on higher-voltage circuits with the help of a relay module.
Credit: Mindsensors

sBrick Control System

The sBrick is a new product that offers the capability to control LEGO robots with a wireless connection (see Figure 9.24). It's plug and play—simply attach it between the battery and the motors. The sBrick's selling points are its range of up to 100 feet, as well as its capability to interface with mobile devices.

While the sBrick can't control Mindstorms motors, it can run Power Functions motors, which can be incorporated into Mindstorms models and use the same sort of gears and axles. Just another way to help make your robots work better for you.

https://www.sbrick.com/

FIGURE 9.24 The sBrick controls LEGO robots wirelessly. Credit: sBrick.

PlayStation Controller Interface

This board plugs into the EV3 and connects wirelessly to a classic PlayStation 2 controller (see Figure 9.25). Sold by Mindsensors, it gives you the ability to move your robot around with joysticks instead of a program.

http://www.mindsensors.com/index.php?module=pagemaster&PAGE_user_op=view_page&PAGE_id=61

FIGURE 9.25 Control a robot with a PlayStation 2 controller. Credit: Mindsensors.

NXTBee Wireless Board

Another existing technology adapted to add value to EV3 are XBee modules, which form mesh networks that allow you to connect potentially dozens of modules together. This lets you control multiple robots from one EV3 brick. Dexter Industries sells a board (see Figure 9.26) that includes a Mindstorms plug and a mounting area for an XBee module (sold separately).

http://www.dexterindustries.com/shop/nxtbee-naked-lego-mindstorms-nxt/

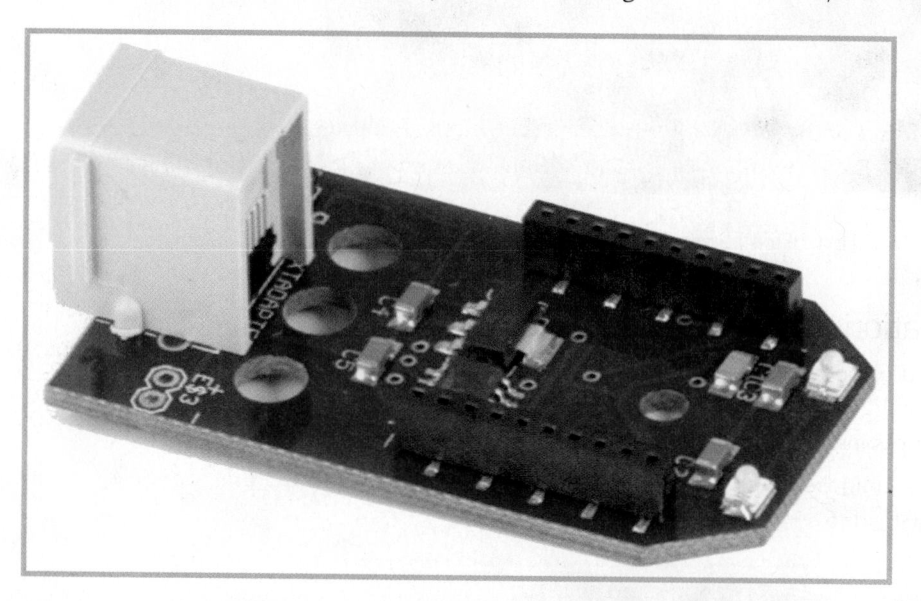

FIGURE 9.26 This board interfaces an XBee wireless module with the EV3 brick. Credit: Dexter Industries.

Summary

In this chapter you checked out a bunch of different add-on electronics modules that give your EV3 brick more capabilities. In Chapter 10, "Project: Flagpole Climber," you put this know-how to the test when you build a robot that climbs a flagpole. Then you add a Raspberry Pi and other robots to give your robot even more capability. Let's go!

10

Project: Flagpole Climber

The final project of the book is the Flagpole Climber, an oddly-shaped robot built to roll up and down poles (see Figure 10.1). After you build the climber you program it in Mindstorms, and then you get to use a number of cool techniques you learned throughout the book to make the climber even better.

First, you add an old-school NXT ultrasonic sensor, a module no longer found in the standard Mindstorms EV3 set. Then you ditch the Intelligent Brick altogether and swap in a Raspberry Pi microcomputer (abbreviated RPi) and a Mindstorms-controlling BrickPi interface board, making the robot exponentially more powerful—at least computationally! Either way it's a great learning experience.

FIGURE 10.1 The Flagpole Climber is a robot that climbs flagpoles.

Building the Flagpole Climber

The most notable thing about the Flagpole Climber might be its shape, with a 30-degree angle in the middle, the thought being to maximize the friction the wheels bear on the pole.

Parts List

Begin by gathering together the following parts (see Figure 10.2):

A. 1x Intelligent Brick

B. 2x motors

C. 4x rims

D. 2x tires

E. 1x tank tread

F. 4x 15M beams

G. 4x 13M beams

H. 4x 11M beams (the red ones)

- **I.** 1x 9M beams
- **J.** 2x 7M beams
- **K.** 4x 5M beams
- **L.** 2x 5x11 chassis bricks
- **M.** 2x 5x7 chassis bricks
- **N.** 3x 3x5 angle beams
- **O.** 2x 4x4 angle beams
- **P.** 2x 3x3 T-beams
- **Q.** 8x 3M beams with pegs
- **R.** 1x 9M cross axle
- **S.** 3x 8M cross axles with end stop
- **T.** 2x 6M cross axles
- **U.** 1x 5M cross axle
- **V.** 53 connector pegs
- **W.** 14 3M connectors
- **X.** 2 cross connectors
- **Y.** 7 bushings
- **Z.** 2 half bushings
- **AA.** 2x Z36 gears
- **BB.** 2x Z12 gears

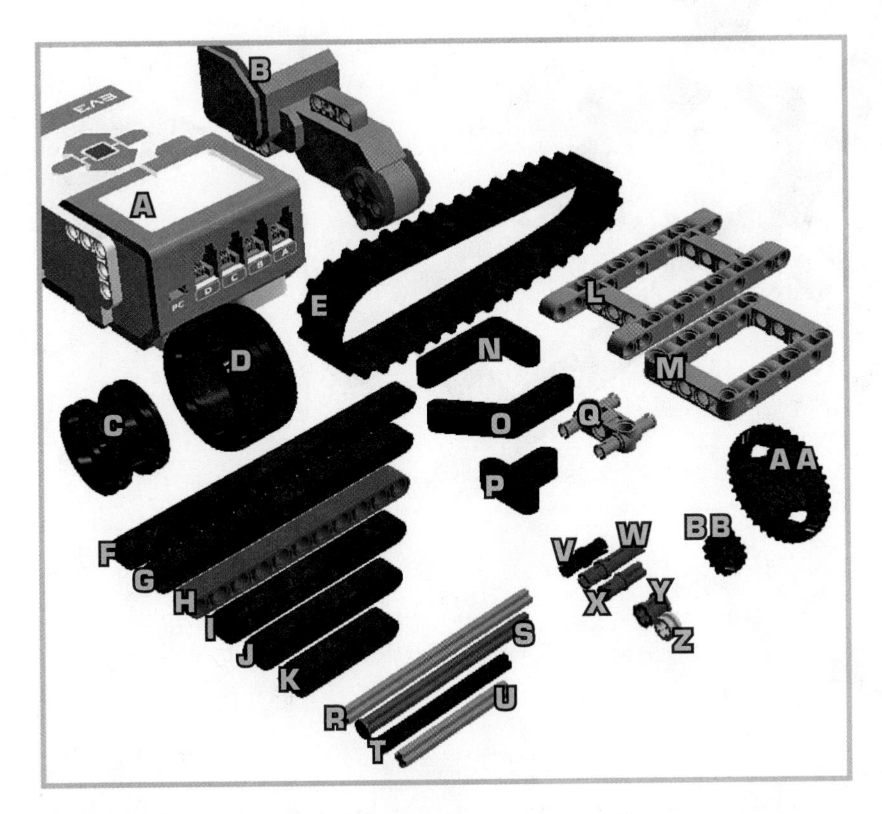

FIGURE 10.2 Gather together these parts to build your climber.

Building Steps

Let's get started building the climber!

STEP 1 Begin by inserting two pegs, two cross connectors, and two 3M pegs into a pair of 5x11 chassis bricks. The angle at which the two bricks come together in Figure 10.3 is how they'll be when attached; you do that in the next step!

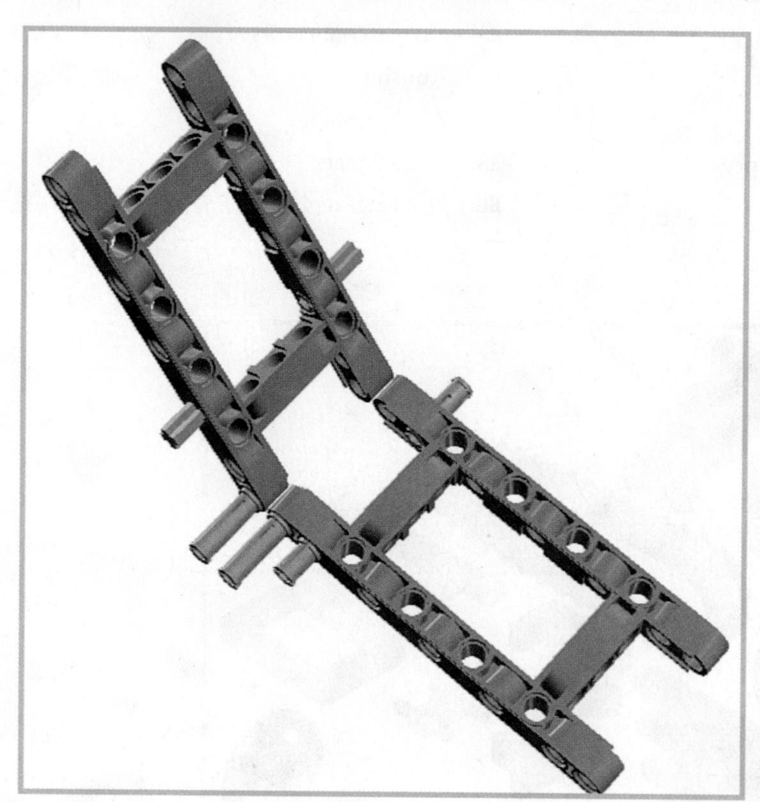

FIGURE 10.3

STEP 2 Secure the two frame bricks using 4x4 angle beams, as shown in Figure 10.4.

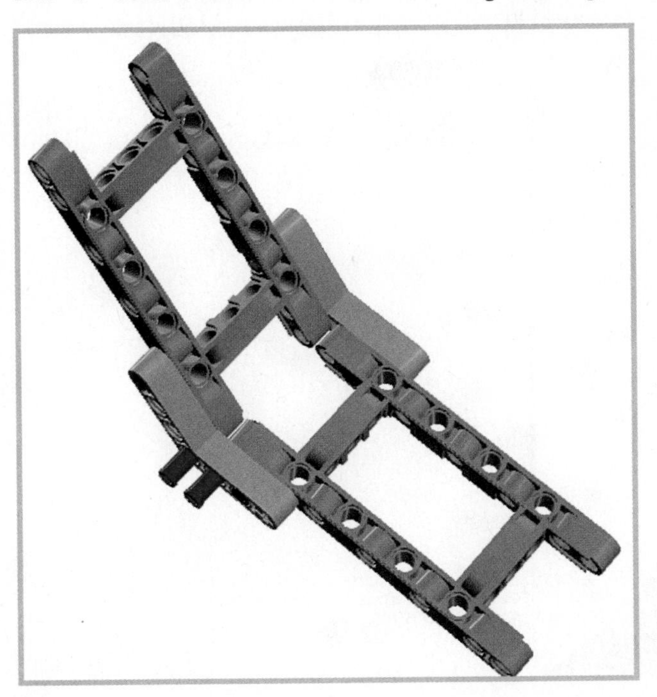

FIGURE 10.4

STEP 3 Add a couple of pegs to the end of the upper frame brick (see Figure 10.5).

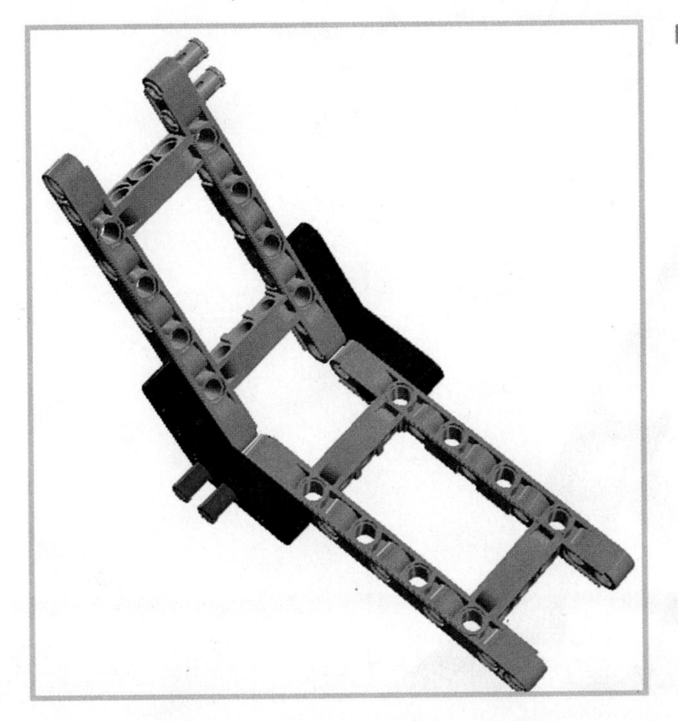

FIGURE 10.5

STEP 4 Attach a 15M beam to the pegs, along with a couple of extra pegs. Figure 10.6 shows how your model should look.

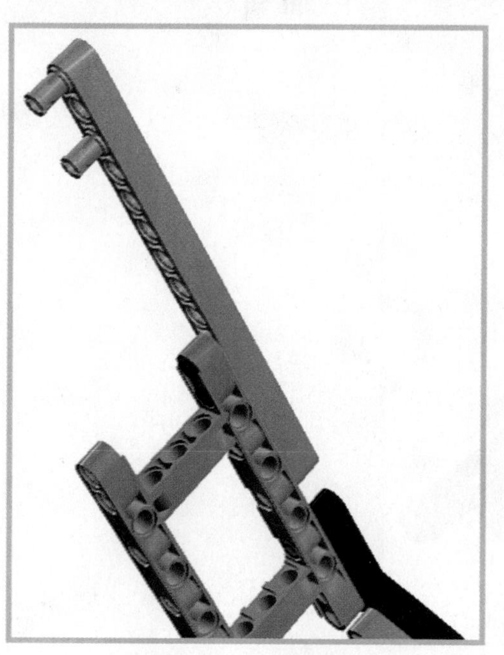

FIGURE 10.6

STEP 5 Add a 5x7 frame brick to the 15M beam, as shown in Figure 10.7.

FIGURE 10.7

STEP 6 Next, insert five pegs into the model (shown in Figure 10.8).

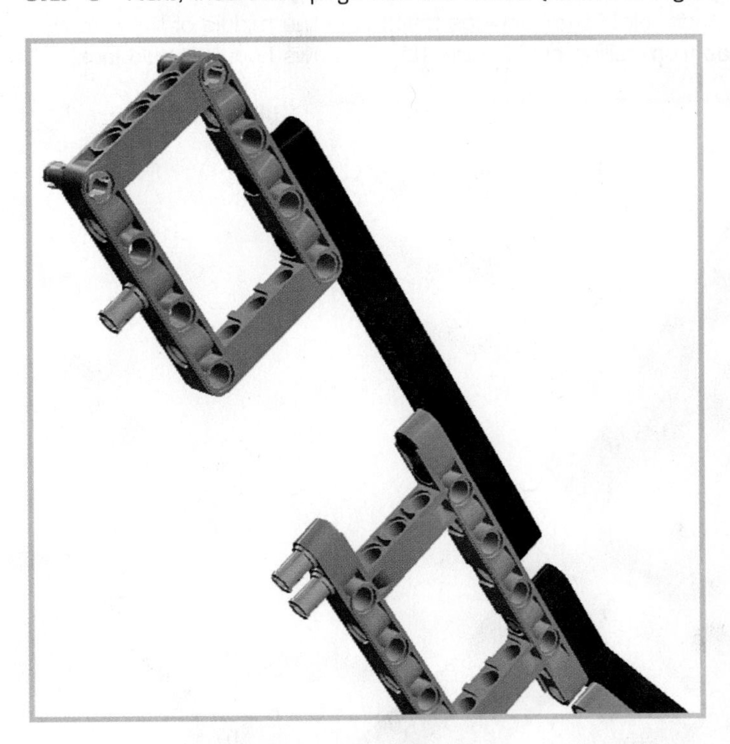

FIGURE 10.8

STEP 7 Attach another 15M beam to match the other (shown in Figure 10.9).

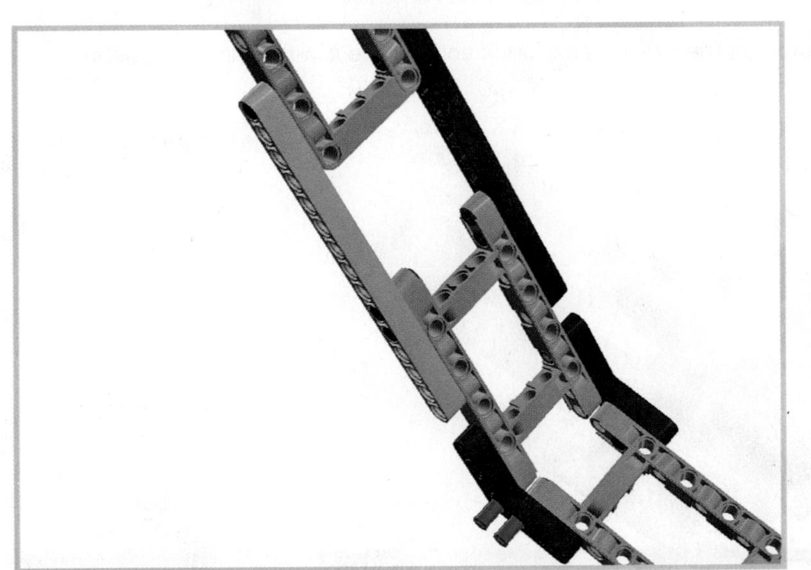

FIGURE 10.9

STEP 8 Insert a motor into the back of the upper frame brick and secure it with a pair of 8M axles with end stops. Note the bushings trapped in the middle of the motor mount! Those keep the axles from falling out. Figure 10.10 shows how it should look.

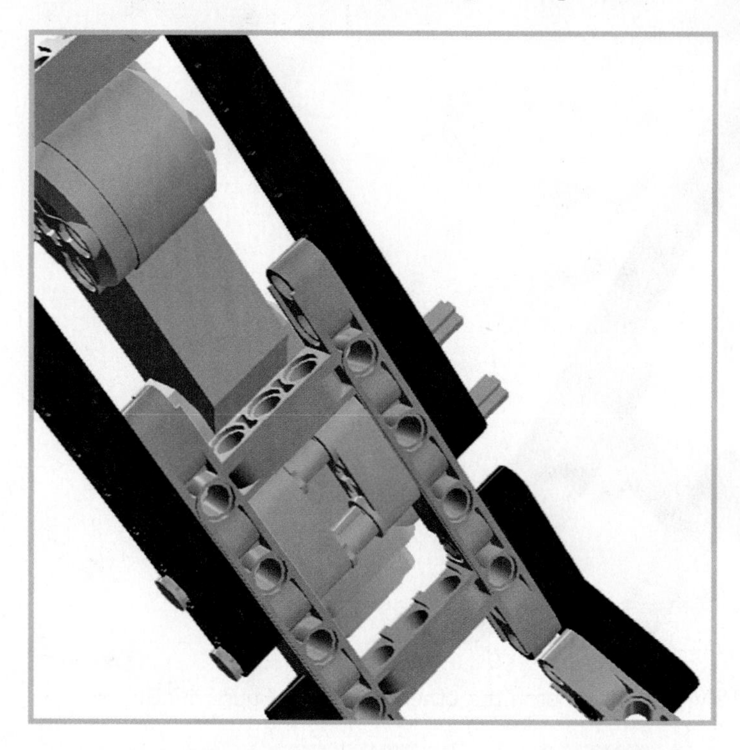

FIGURE 10.10

STEP 9. Put a motor on the lower frame brick and secure it with four 3M pegs, as shown in Figure 10.11.

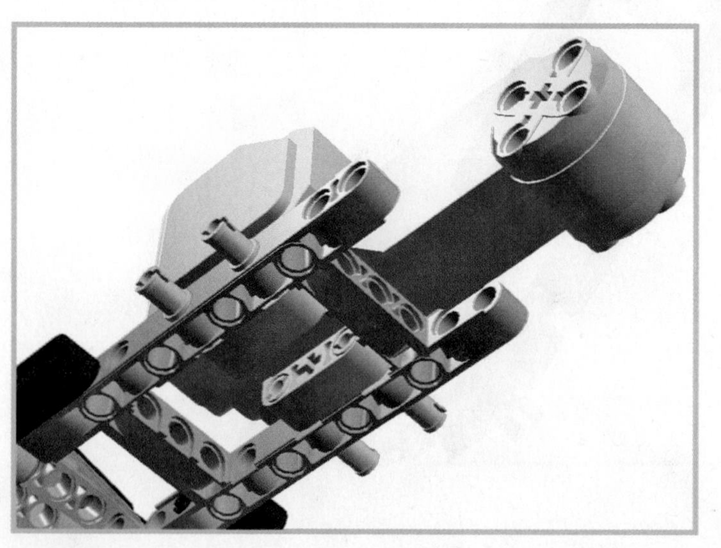

FIGURE 10.11

STEP 10. Insert a 6M axle and bushing into the hub of the lower motor (shown in Figure 10.12).

FIGURE 10.12

STEP 11 Place a 13M beam onto the exposed pegs, as well as the axle you just added (see Figure 10.13).

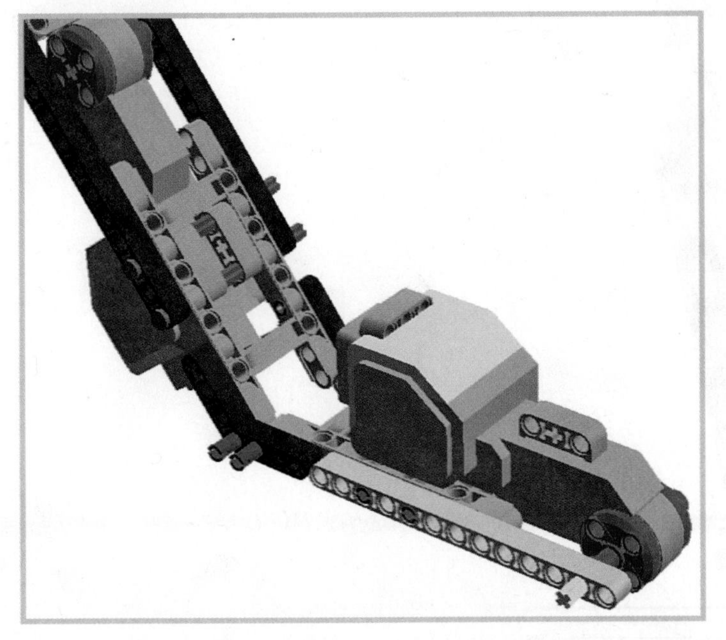

FIGURE 10.13

STEP 12 Add a couple more pegs, as shown in Figure 10.14.

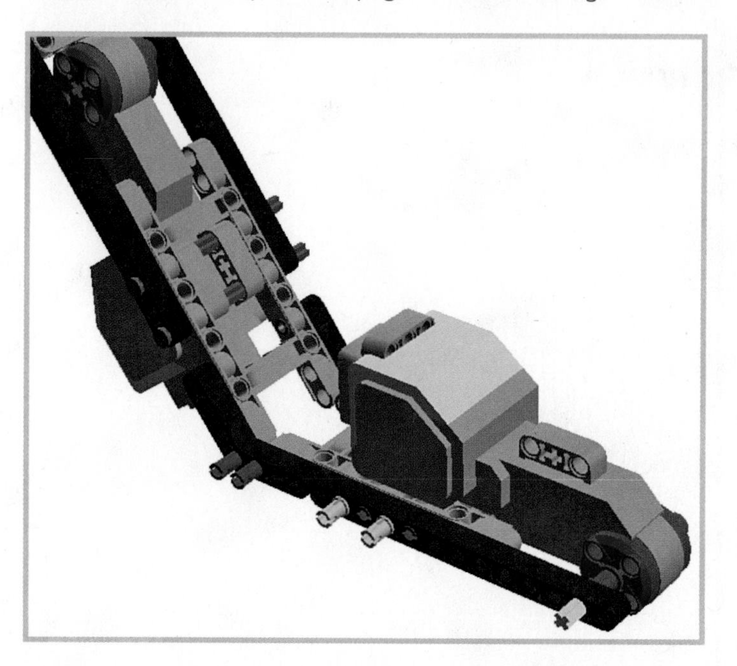

FIGURE 10.14

STEP 13 Insert a 8M axle with end stop into the hub of the upper motor, trapping a pair of bushings (one on each side of the motor) along the way (see Figure 10.15).

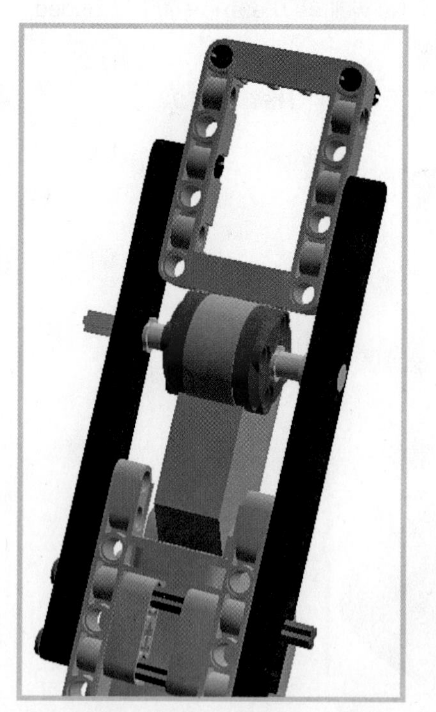

FIGURE 10.15

STEP 14 Add Z12 gears to both axles, as shown in Figure 10.16.

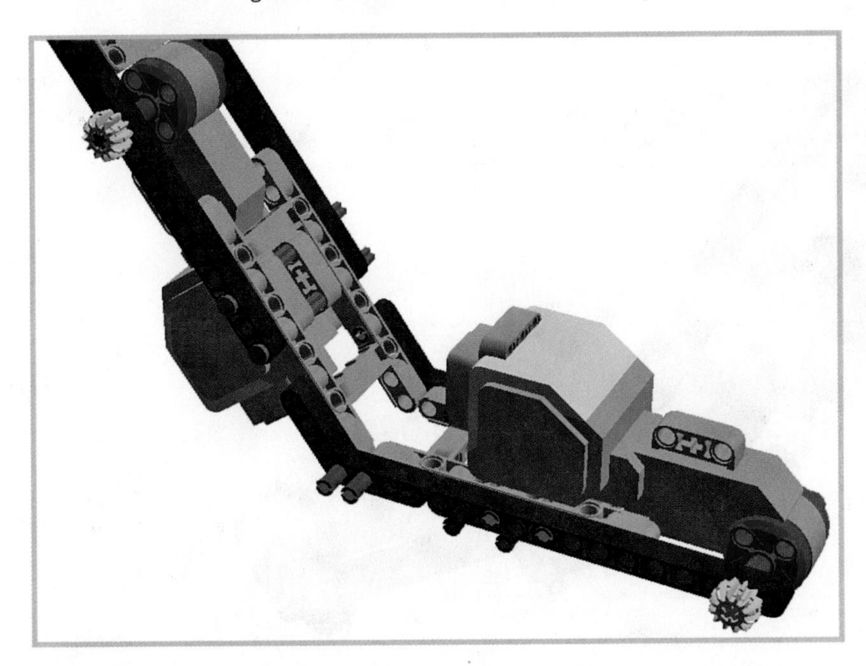

FIGURE 10.16

STEP 15 Add an 11M beam to the pegs (shown in Figure 10.17).

FIGURE 10.17

STEP 16 Add a pair of 3M beams with pegs, as shown in Figure 10.18.

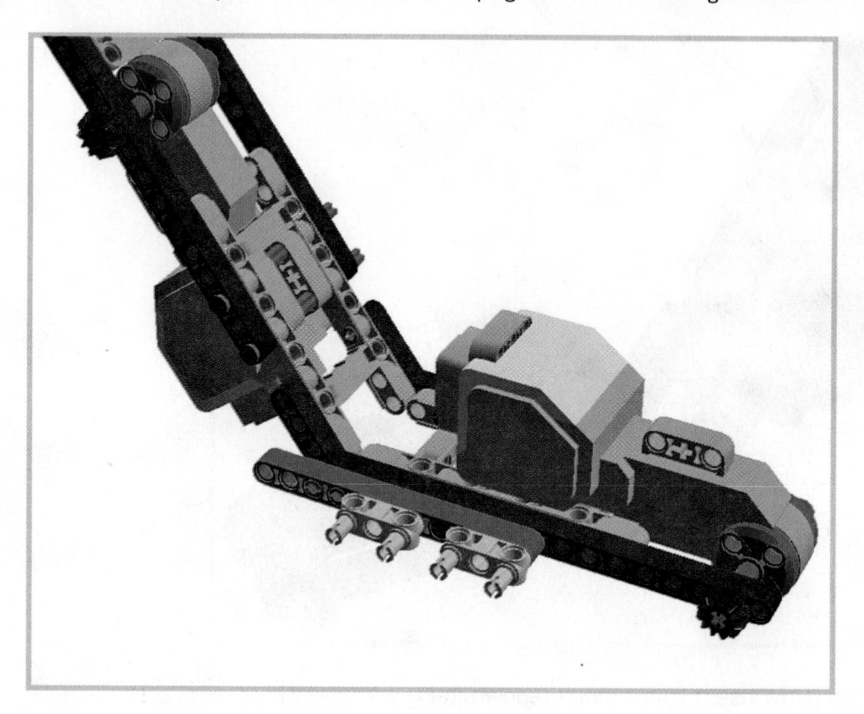

FIGURE 10.18

STEP 16 Attach a 7M beam, two pegs, and a 3M peg. Figure 10.19 shows how the climber should look.

FIGURE 10.19

STEP 18 Attach a 5M beam, as shown in Figure 10.20.

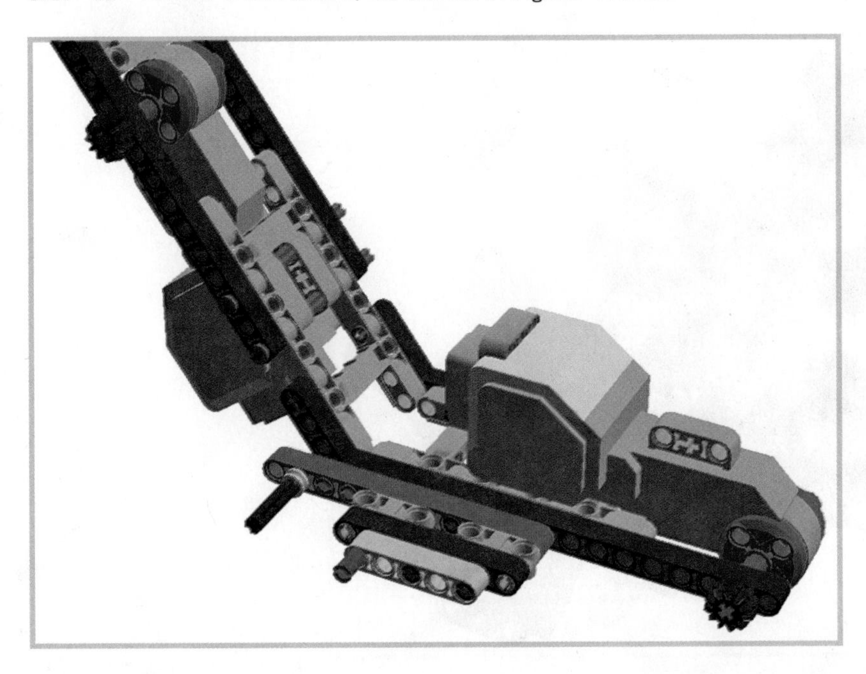

FIGURE 10.20

STEP 19 Attach a pair of 3M beams with pegs to the bottom of the frame brick, as shown in Figure 10.21.

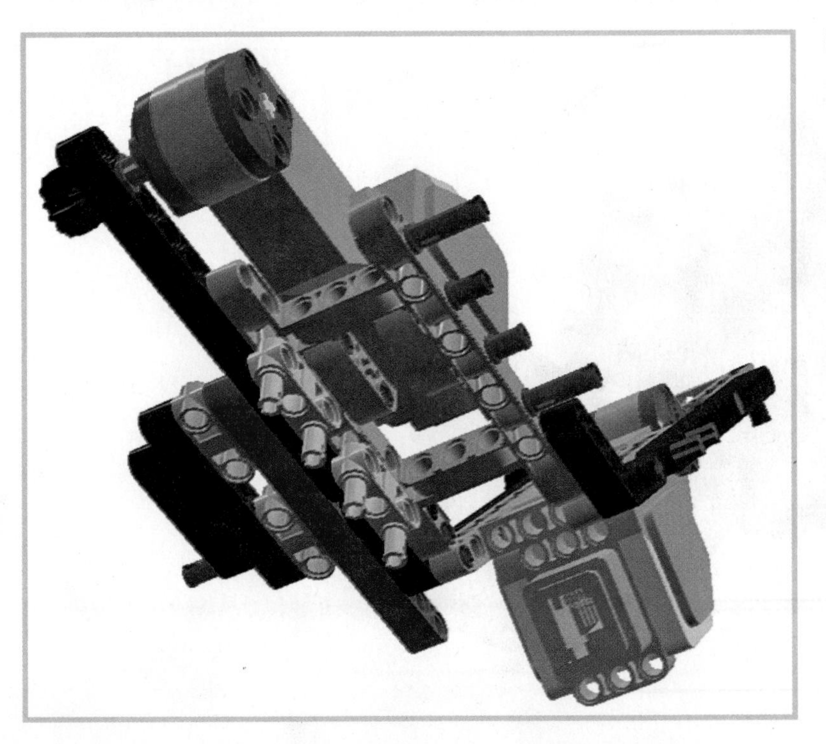

FIGURE 10.21

STEP 20 Add a 13M beam. Figure 10.22 shows how it should look.

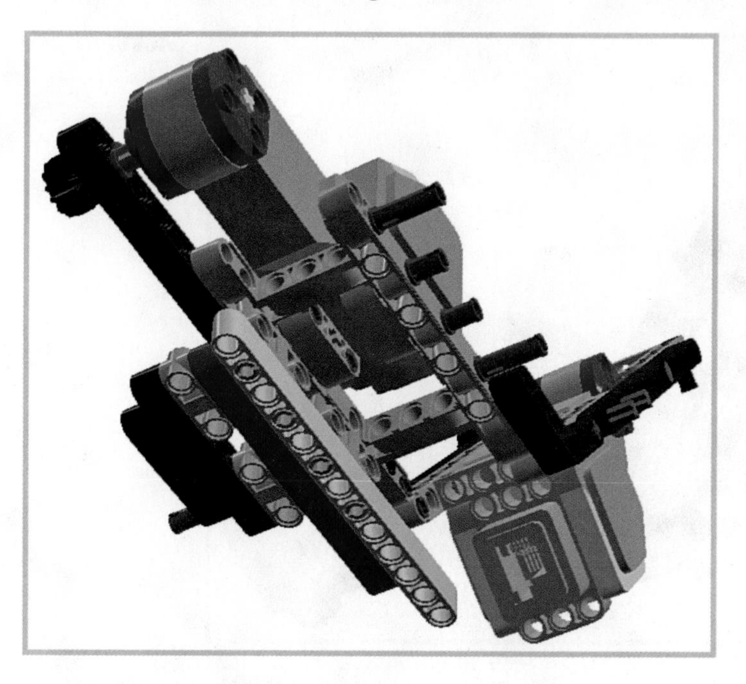

FIGURE 10.22

STEP 21 Insert two normal pegs and a 3M peg to the 13M beam you just added (shown in Figure 10.23).

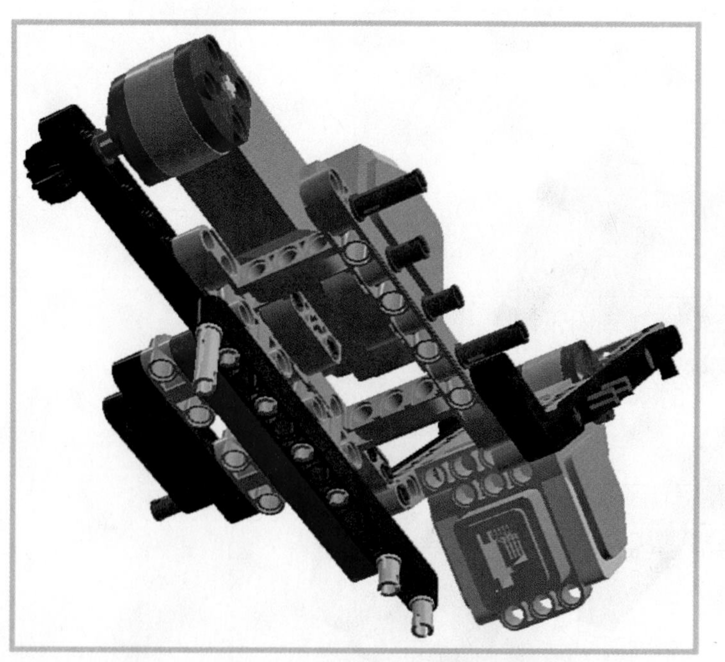

FIGURE 10.23

STEP 22 Add a pair of 3x5 angle beams to the underside of the climber, as shown in Figure 10.24.

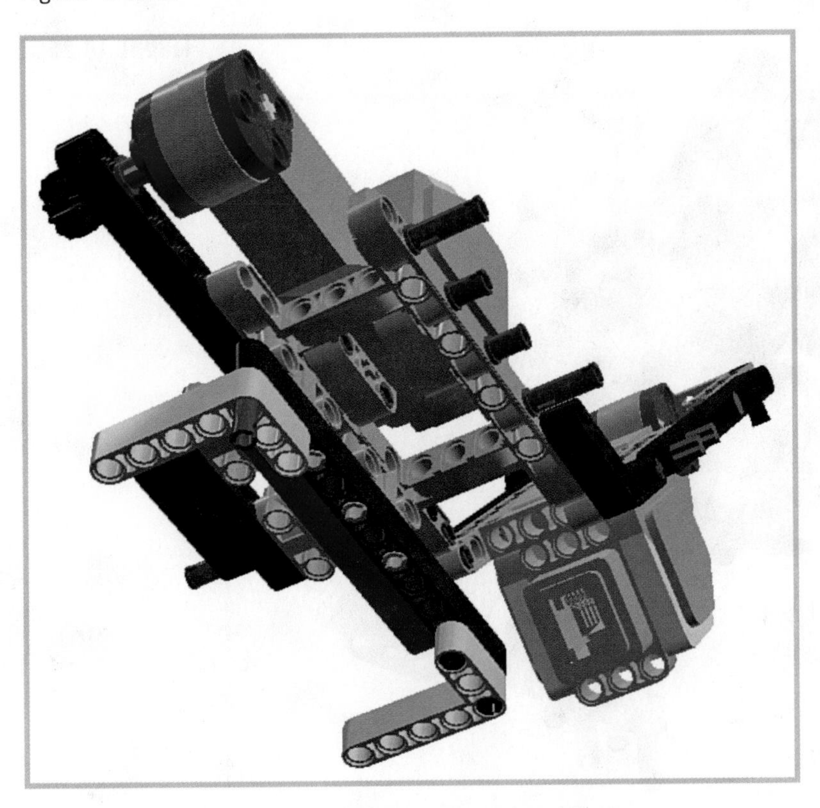

FIGURE 10.24

STEP 23 Insert three normal and four 3M pegs to the angle beams, as shown in Figure 10.25.

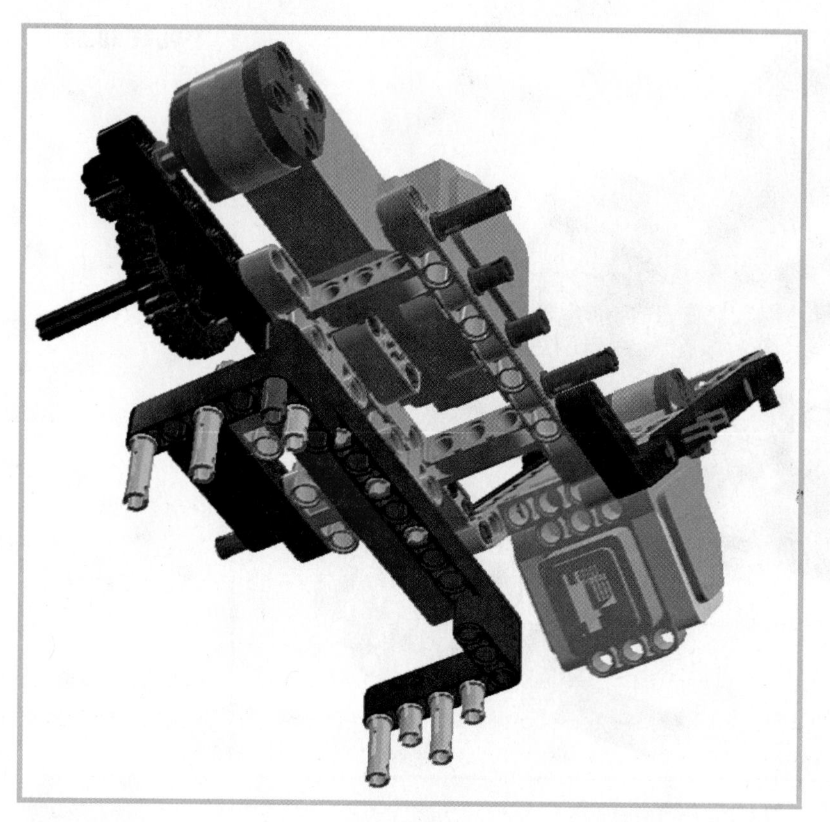

FIGURE 10.25

STEP 24 Add a pair of 5M beams. Figure 10.26 shows how it should look.

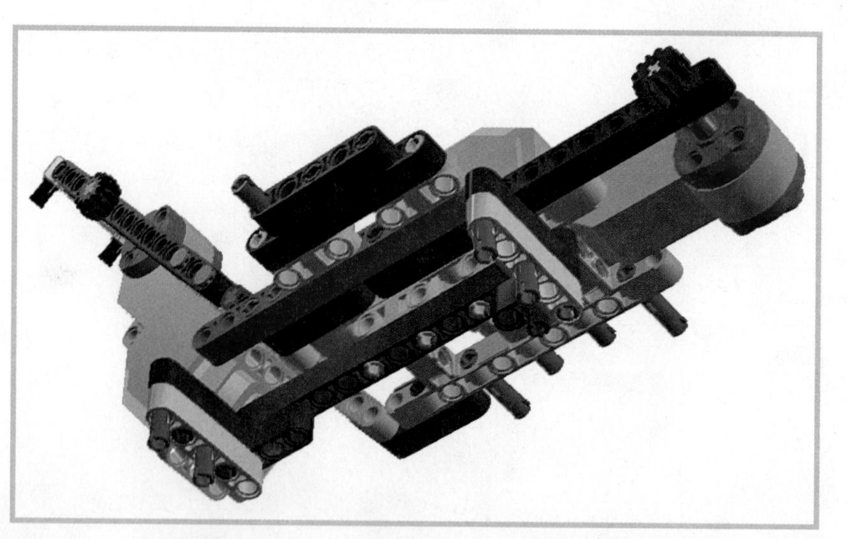

FIGURE 10.26

STEP 25 Insert a 6M axle and a Z36 gear such that the gear meshes with the smaller Z12 gear (as shown in Figure 10.27).

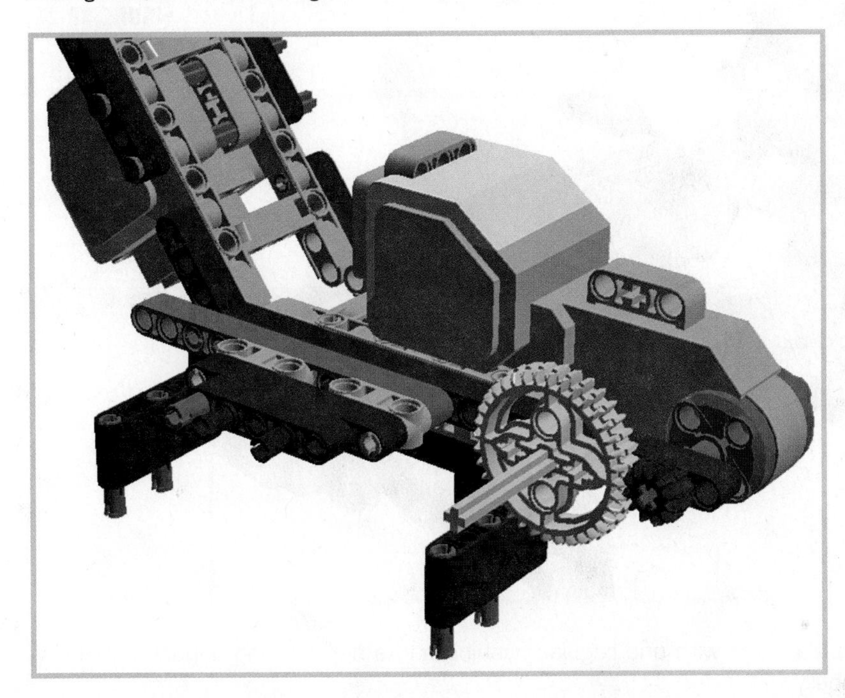

FIGURE 10.27

STEP 26 Insert a 6M axle into the red beam and add a half bushing. This will want to fall out, but just hold it in place with your finger (see Figure 10.28).

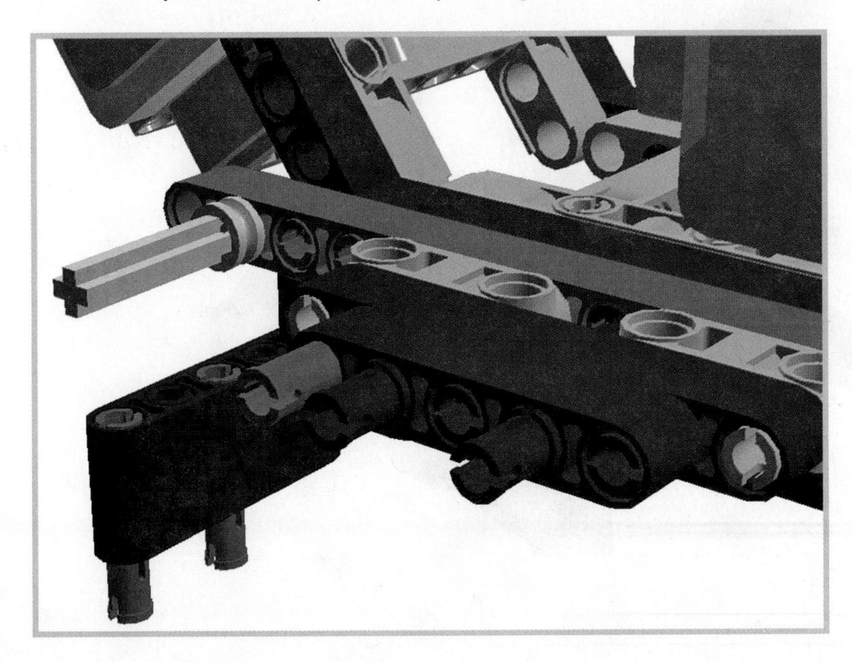

FIGURE 10.28

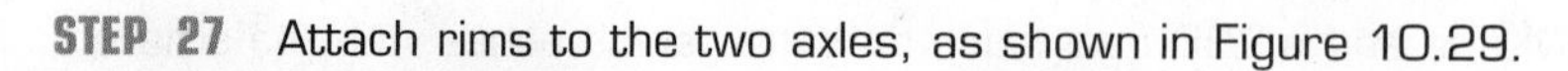

STEP 27 Attach rims to the two axles, as shown in Figure 10.29.

FIGURE 10.29

STEP 28 Secure the rims with one regular bushing and a half bushing. Figure 10.30 shows how it should look.

FIGURE 10.30

STEP 29 Put the tank treads on the robot. My design program doesn't do treads very well, so here's a photo that shows the tank treads on the finished robot (see Figure 10.31). If your treads don't want to fit in, stretch it out a bit with your fingers, being careful not to pull too hard.

FIGURE 10.31

STEP 30 Add 5M and 9M beams to help stabilize the axles (see Figure 10.32).

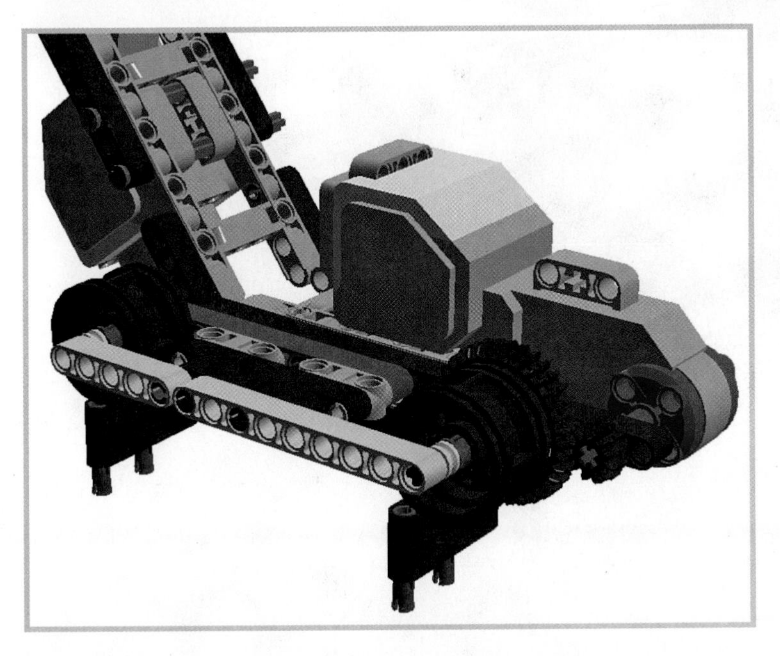

FIGURE 10.32

STEP 31 Place six pegs on these new beams, spacing them as you see in Figure 10.33.

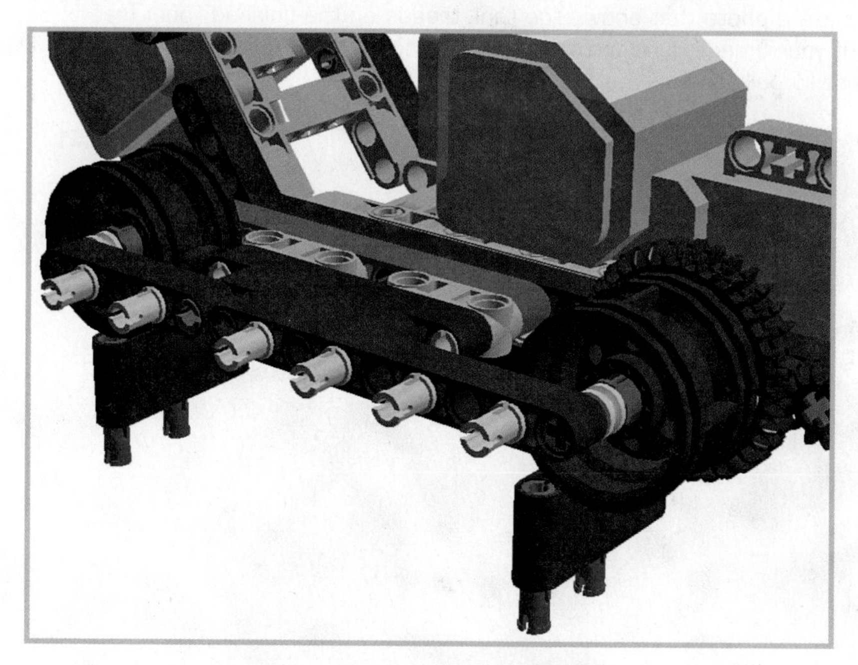

FIGURE 10.33

STEP 32 Throw on a 13M beam, shown in Figure 10.34.

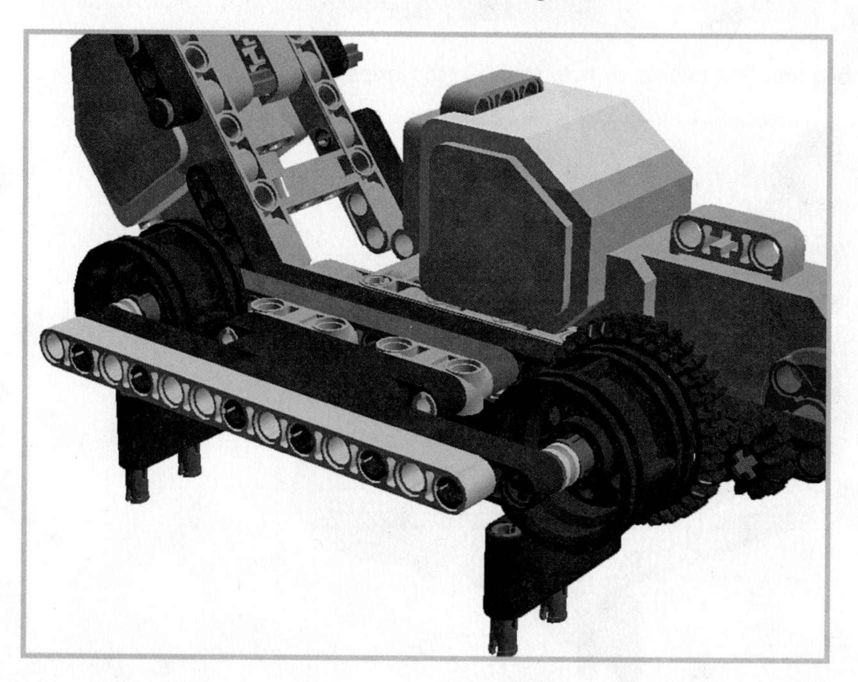

FIGURE 10.34

STEP 33 Add three more pegs to the 13M beam, shown in Figure 10.35.

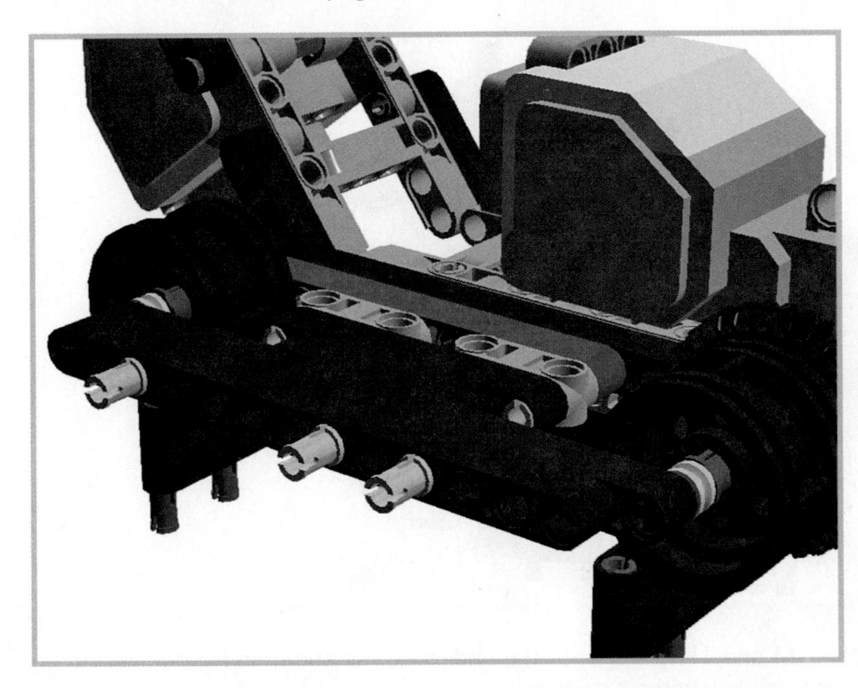

FIGURE 10.35

STEP 34 Attach the Intelligent Brick, as shown in Figure 10.36. Held on by only three pegs, the Brick is going to wobble, so keep it steady until it can be secured.

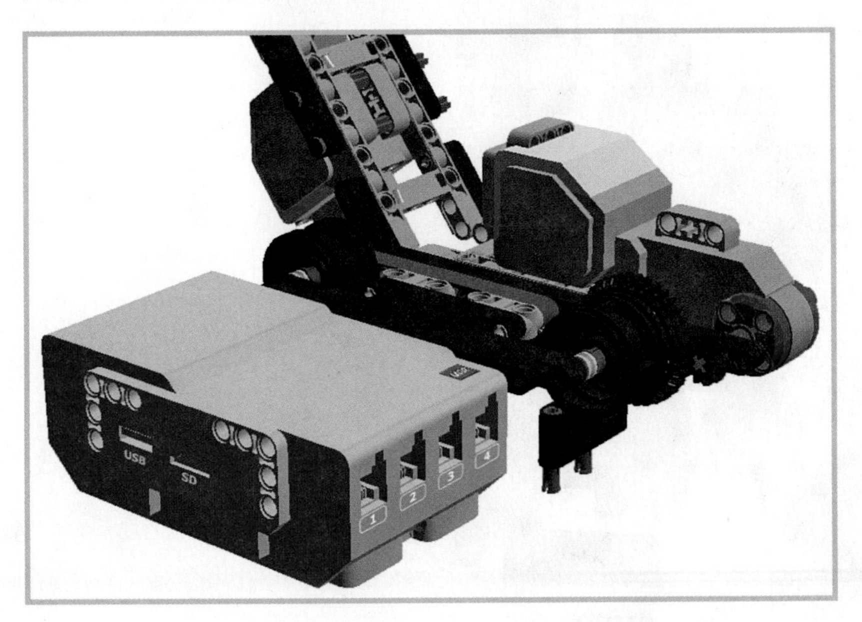

FIGURE 10.36

STEP 35 Add six pegs to the underside of the Intelligent Brick (see Figure 10.37).

FIGURE 10.37

STEP 36 Attach 11M and 13M beams to the underside, as shown in Figure 10.38. Now the Intelligent Brick has some proper support!

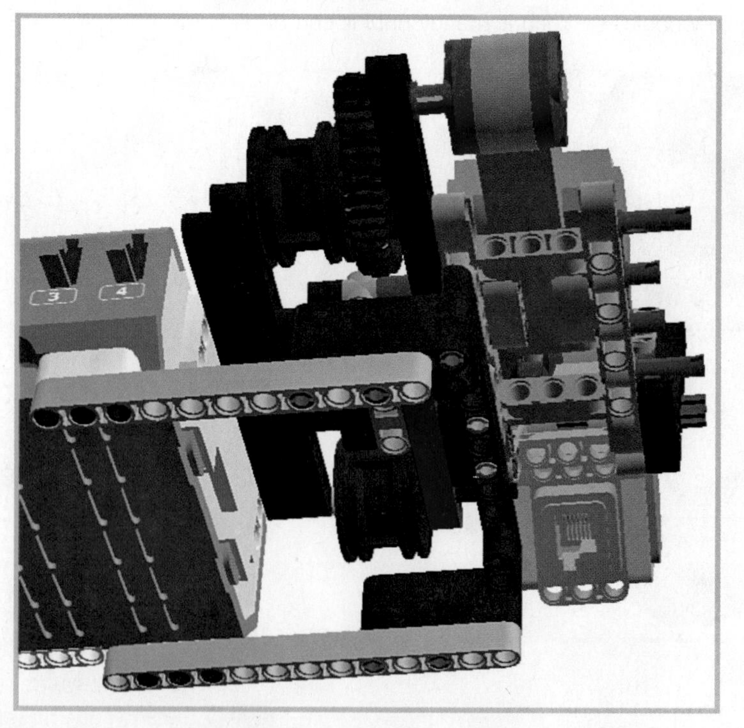

FIGURE 10.38

STEP 37 Add a 7M beam to the exposed pegs on the other side. Figure 10.39 shows how your robot should look.

FIGURE 10.39

STEP 38 Continuing on the same side, add six connector pegs as shown in Figure 10.40.

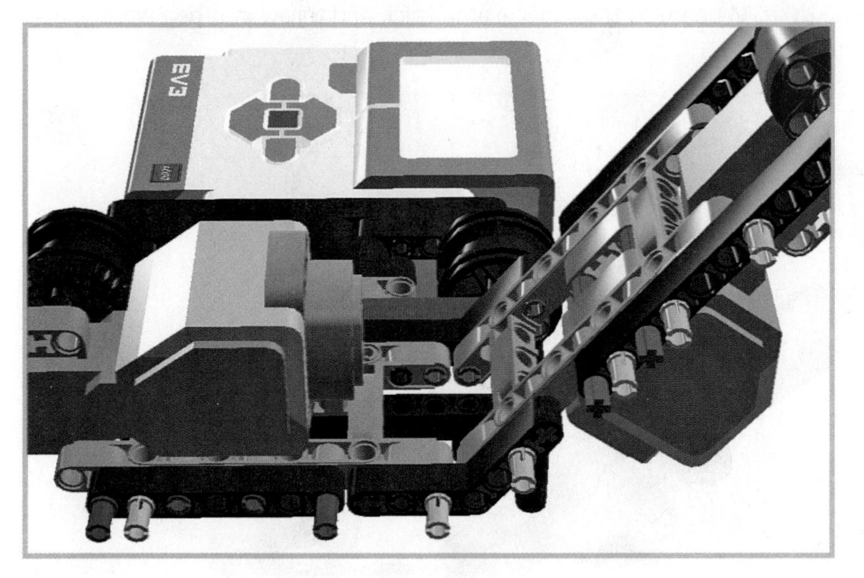

FIGURE 10.40

STEP 39 Attach a pair of 11M beams to the side of the climber (shown in Figure 10.41).

FIGURE 10.41

STEP 40 Connect a 15M beam to the other frame brick and throw in a peg while you're at it (see Figure 10.42).

FIGURE 10.42

STEP 41 Let's switch gears a moment (ha!) and work on the upper wheel assembly. Begin by adding a Z36 gear to a 7M axle, as shown in Figure 10.43.

FIGURE 10.43

STEP 42 Slide two rims onto the axle (shown in Figure 10.44).

FIGURE 10.44

STEP 43 Secure the rims with two bushings. The assembly should look like Figure 10.45.

FIGURE 10.45

STEP 44 Connect one end of the wheel assembly so that the Z36 gear meshes with the Z12 gear attached to the motor. Trap the assembly with the help of a 5x7 chassis brick, which is attached to the 15M beam you added in Step 40. It should look like Figure 10.46.

FIGURE 10.46

STEP 45 Add three pegs to the underside of the chassis brick, as shown in Figure 10.47.

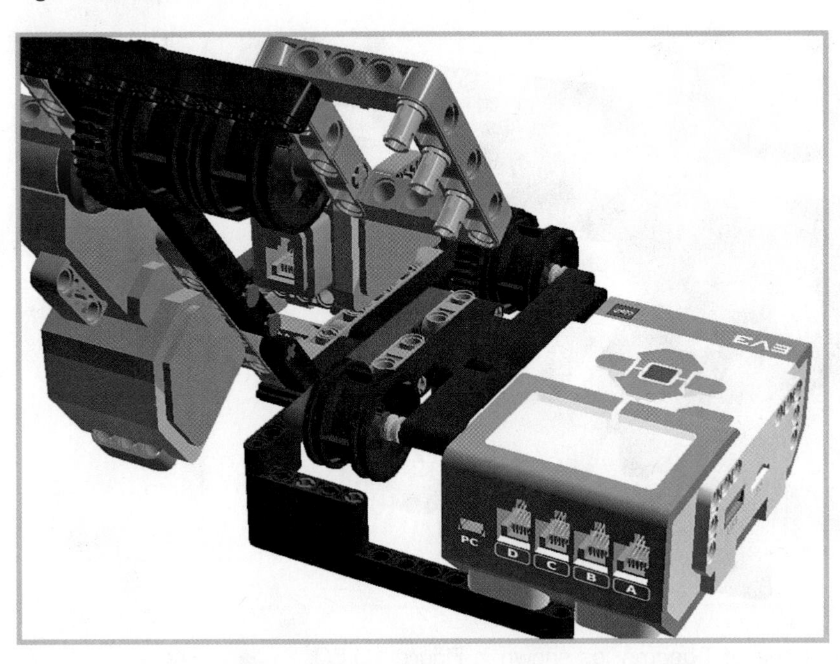

FIGURE 10.47

STEP 46 Attach a 3x5 angle beam to the pegs you just added (see Figure 10.48).

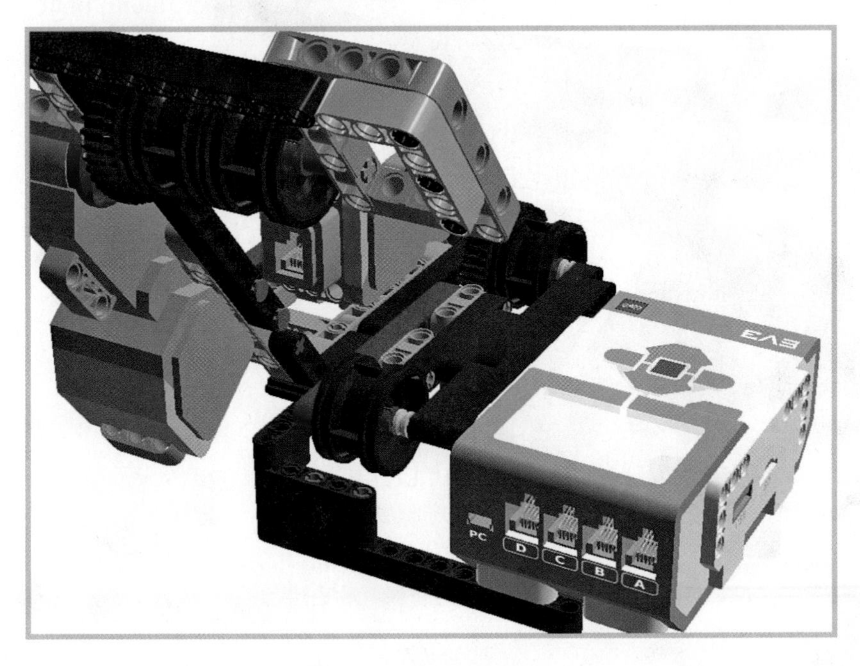

FIGURE 10.48

STEP 47 Add a pair of 3M beams with pegs to the undersides of the two 5x7 chassis bricks, as shown in Figure 10.49.

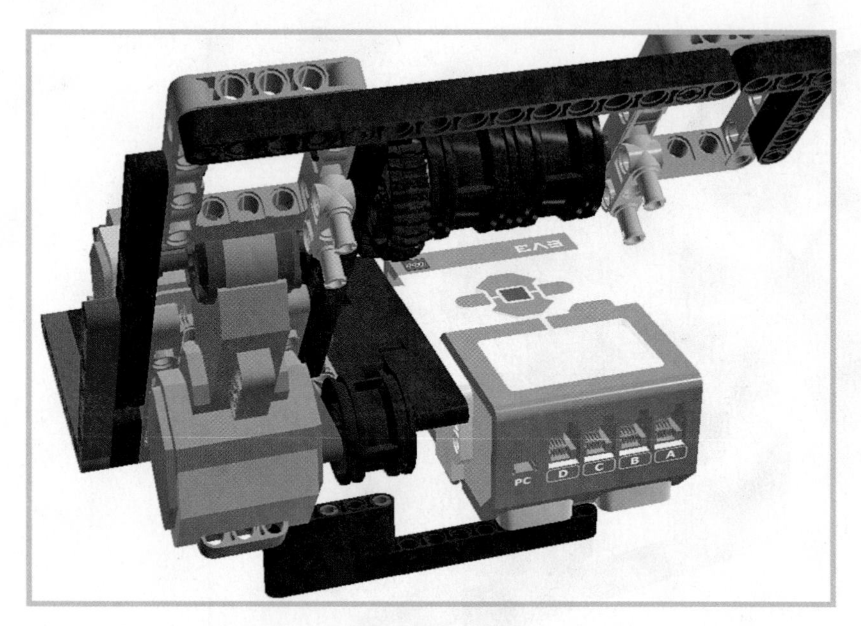

FIGURE 10.49

STEP 48 Attach a pair of T-beams, as shown in Figure 10.50.

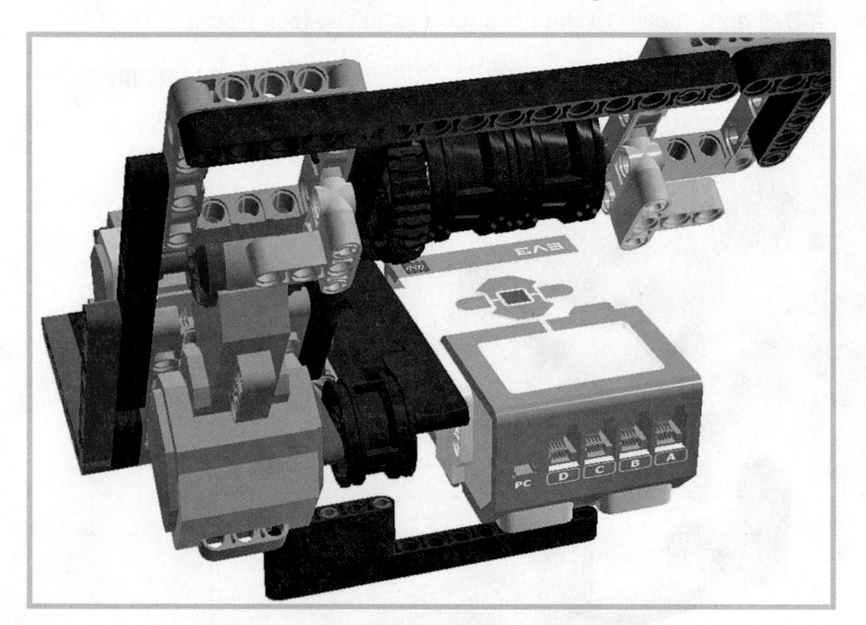

FIGURE 10.50

STEP 49 Attach another pair of 3M beams with pegs. Figure 10.51 shows how it should look.

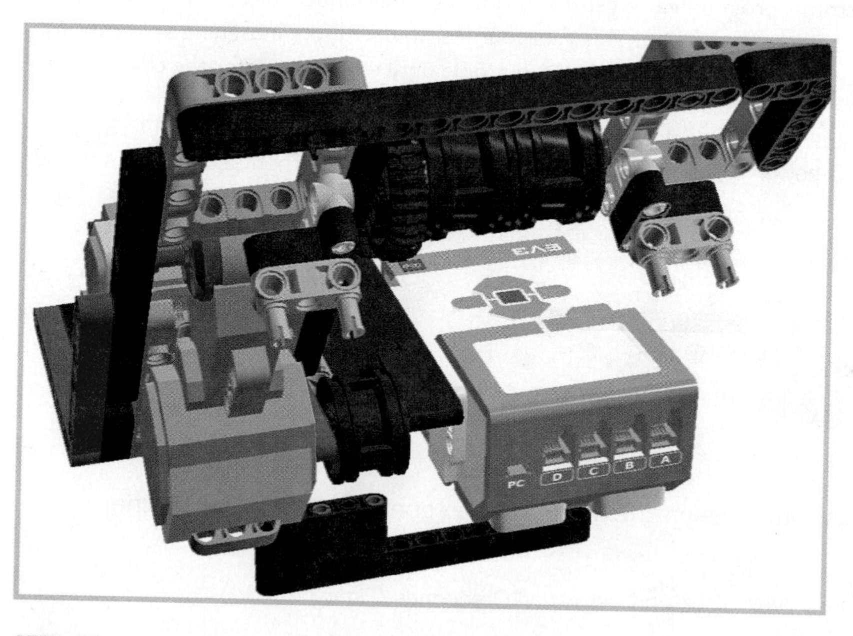

FIGURE 10.51

STEP 50 Finish up the robot's basic build by adding a 15M beam, as shown in Figure 10.52.

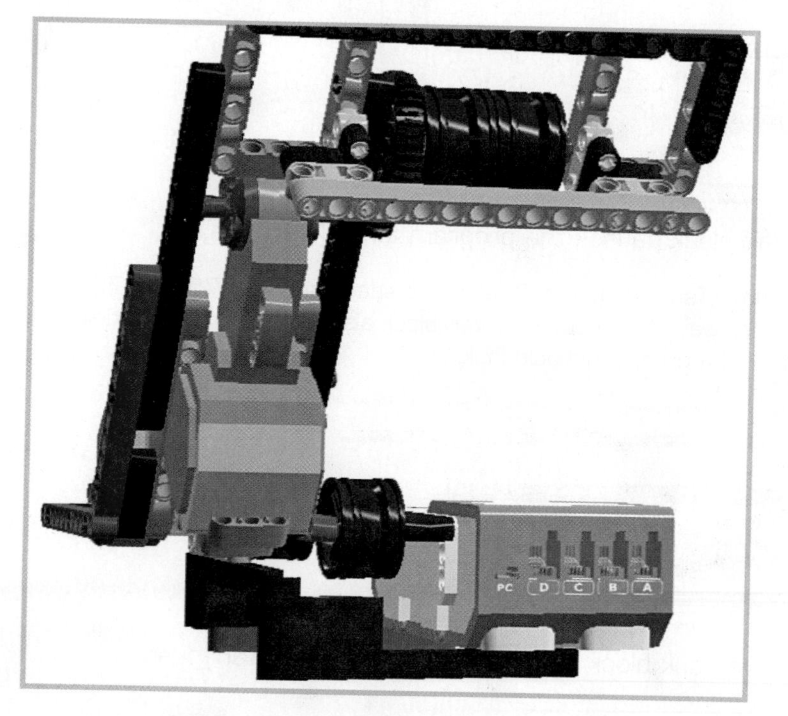

FIGURE 10.52

Programming the Pole Climber

The basic Mindstorms program can't get much simpler: The climber rolls up the pole a short distance and then back down again. This gets your feet wet for more intriguing applications, and this way you won't get your robot stuck up a pole until you're ready for the challenge!

STEP 1 Begin with a Loop block and drop in a Move Tank block (see Figure 10.53). Set one motor to -75 power and leave the other at its default of 75. Then set rotations to 10.

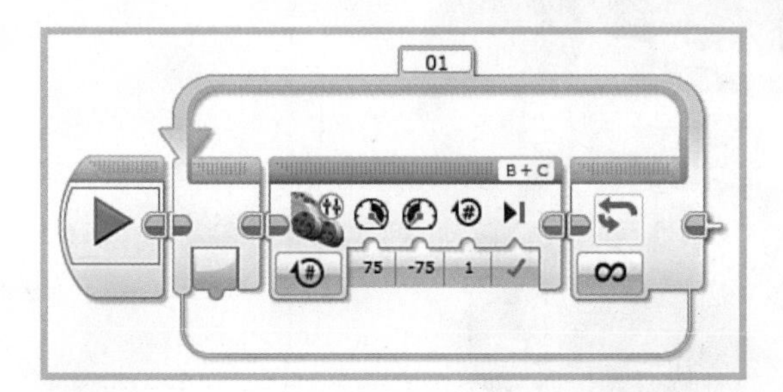

FIGURE 10.53 The loop keeps the instructions repeating until you halt the program.

STEP 2 Drop in a Wait block and set it to 10 seconds (Figure 10.54).

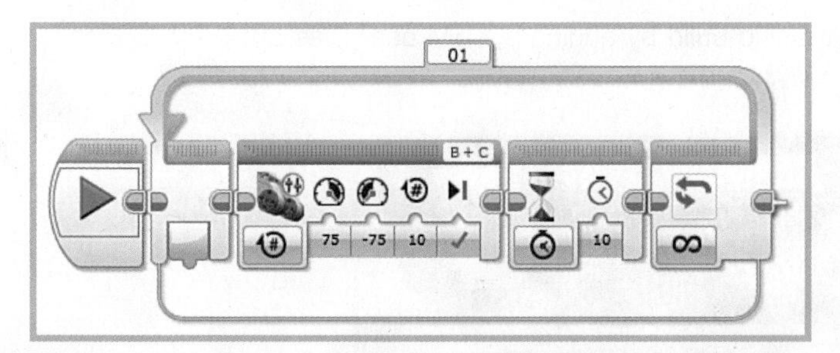

FIGURE 10.54 The Wait block pauses the program for 10 seconds.

STEP 3 Add another Move Tank block, but reverse the speed settings (75 and -75) so that it will roll the other way. Add a second Wait block also set to 10 second delay. Figure 10.55 shows how the program should look.

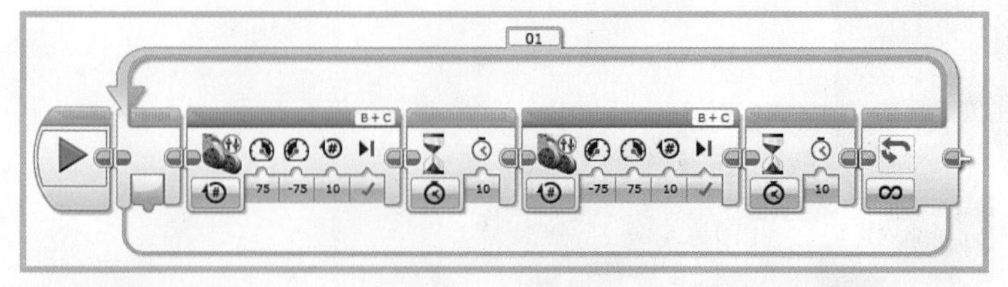

FIGURE 10.55 The Move Tank block controls two motors simultaneously.

Installing the Ultrasonic Sensor

As mentioned, the previous version of Mindstorms (NXT) included an ultrasonic sensor, a gadget that sends out inaudible "pings" consisting of sound waves and measures how fast they bounce back, thereby determining the distance to an obstruction. Here's how to add this gadget to your robot:

STEP 1 Buy a sensor: You can still purchase them from LEGO, or they may be found at yard sales, on auction sites, or online retailers. http://shop.lego.com/en-US/Ultrasonic-Sensor-9846

STEP 2. Download the programming block: You can also find it on LEGO, on the Mindstorms downloads page at http://www.lego.com/en-us/mindstorms/downloads

STEP 3 From the EV3 software, select Tools and then Block Import. Browse to the folder where you put the block and then hit import. You need to reboot the software to see the new block.

STEP 4 Attach the sensor to the robot, using a 3x7 angle beam as pictured in Figure 10.56. This positions the sensor in a way that it's pointing down when the robot's climbing a pole. Connect the sensor to port 4 on the Intelligent Brick.

FIGURE 10.56 Attach the ultrasonic sensor to the robot.

Programming the Sensor

I decided to have a Switch block provide the main control mechanism for the ultrasonic sensor. The Switch moves the robot upward until the robot reaches 240cm (around 8 feet) in altitude. When that happens, the robot changes direction and rolls down until it is within 10cm of the ground, or around 4 inches. It's a basic program that whets our appetite for more complexity!

1. Drag out a Switch block from the Flow Control palette and configure it to be triggered by the ultrasonic sensor, set to Compare and Measure Centimeters. Set it to trigger when the distance measured is less than or equal to 240cm.
2. Put a Move Tank block inside the top cell, as shown in Figure 10.57. Configure the Move Tank block as you did in the previous section. Also throw in a Wait block while you're there, configured as before. The lower cell of the Switch remains empty.

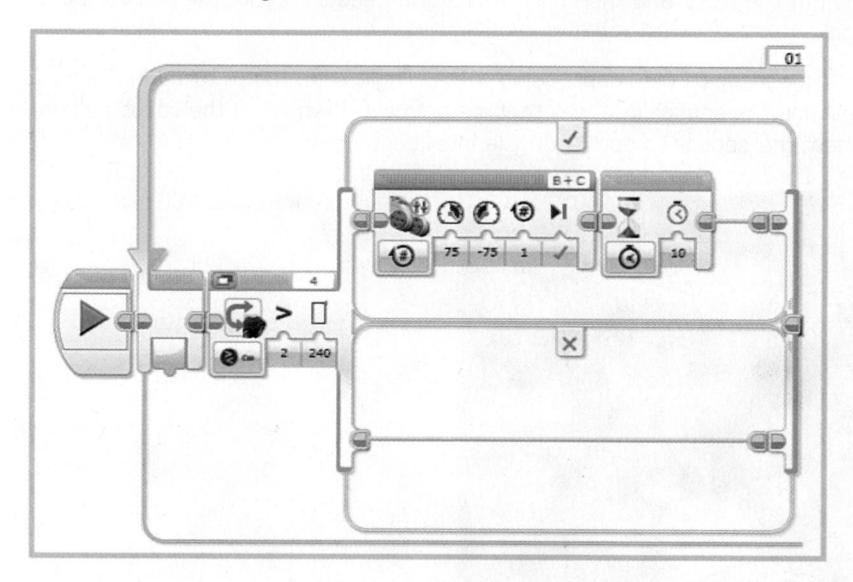

FIGURE 10.57 Use a Switch block to control the robot.

3. After the main Switch, drop in another Switch, as shown in Figure 10.58. The first Switch runs the motors until the distance to the ground reaches 240cm, or about 8 feet. When that height is reached, the EV3 looks in the bottom cell to see what to do, and sees nothing: That Switch is done. The program moves on to the next Switch, which works the opposite way, rolling the robot downward until it reaches 10cm.

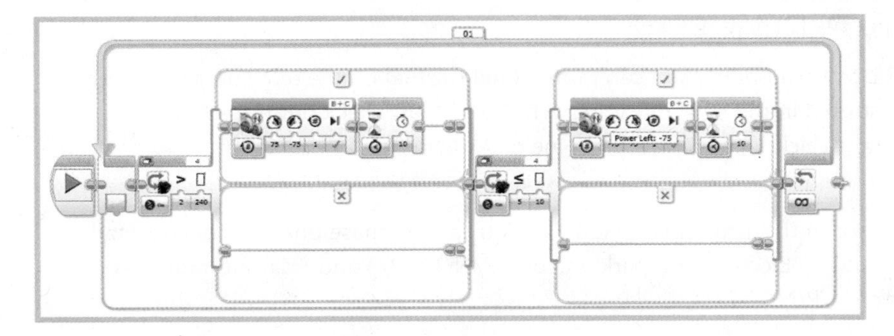

FIGURE 10.58 A second Switch sends the robot back toward earth.

Swapping in the BrickPi

Suppose we swapped out the Intelligent Brick for another controller. In this case I'm talking about the Raspberry Pi microcomputer (see Figure 10.59), a board the size of a credit card that runs the Linux operating system. It can be controlled much the same way as an Arduino but is much more powerful. It can connect to the Internet, play movies, and more. You will learn how to use the RPi in conjunction with a BrickPi board to control the climber. First, however, I will briefly bring you up to speed on the Raspberry Pi.

FIGURE 10.59 The Raspberry Pi microcomputer controls robots and a whole lot more! Credit: raspberrypi.org

Raspberry Pi Quick Start

This isn't a Raspberry Pi book, so I can't offer a full tutorial on the technology. However, I can get you started in the right direction. When in doubt, check out https://www.raspberrypi.org/, which is the product's home organization.

1. Buy an RPi.

   The RPi I used in this project is a Model B. You can purchase one from online retailers such as SparkFun Electronics (sparkfun.com, P/N 13297) and Adafruit Industries (Adafruit.com, P/N 2358).

2. Equip it.

   You can buy accessories for the RPi from the same stores. You need a keyboard, mouse, and display to interface with the unit. In addition, you may want to buy sensors, switches, and other electronics to add on. Finally, many third-party boards—such as the BrickPi are available to add display options, motor control capability, and sensors onto the microcomputer. I get more into what you need to buy in the parts list that follows.

3. Load the OS.

   Another thing to buy is a good SD card, upon which you load the operating system—though many sites offer pre-imaged SD cards for sale. Either way, the RPi's operating system is resident on the card.

4. Program it.

   There is actually a dizzying array of RPi programming options. It is, after all, effectively a small computer. Because it's a computer, uploading a program works about the same way. How do you get a file onto your laptop? Plug in a flash drive, upload it over the network, or put it on the SD card. For this book I chose Python, a relatively simple programming language, to control the Mindstorms robot. You can learn more about Python at python.org.

5. Learn more.

   Dexter Industries has a complete tutorial on installation (http://www.dexterindustries.com/BrickPi/getting-started/). In addition, SparkFun offers an exhaustive tutorial on installing the BrickPi that can be referenced on SparkFun.com: https://learn.sparkfun.com/tutorials/getting-started-with-the-brickpi.

Adding the RPi and BrickPi

Let's do away with that trusty but limited Intelligent Brick and replace it with the RPi and a BrickPi interface board, along with a case to enclose everything. You can see the assembly pictured in Figure 10.60. The BrickPi interfaces between the RPi and Mindstorms motors and features sensor and motor plugs compatible with Mindstorms wires. So basically, it is similar to the Bricktronics Shield mentioned in Chapter 6, "Project: Robot Flower," except for a different platform.

In addition to swapping in the RPi, you take advantage of the BrickPi's Mindstorms compatibility to continue to run the ultrasonic sensor, and you wire in a combo barometric sensor, temperature sensor, and altimeter that gives the RPi more detailed information on what life at the top of a flagpole is like.

FIGURE 10.60 The BrickPi helps the RPi interface with Mindstorms parts.

Parts List

Let's go over what you need to rebuild your climber as a computer-controlled weather station!

- The Mindstorms robot you built earlier in the chapter, including the ultrasonic sensor.
- Raspberry Pi Model B: The latest and greatest, at least at the time of this writing.
- BrickPi: Buy it from dexterindustries.com.
- MPL3115A2 Altimeter: Found at Adafruit.com, P/N 1893.
- Female-to-male jumper wires: Adafruit.com, P/N 1954.
- The BrickPi acrylic case, http://www.dexterindustries.com/shop/brickpi-b-case/.
- Battery pack, such as the one at http://www.dexterindustries.com/shop/brickpi-power-pack/.

Steps

Follow these steps to add the Pi:

STEP 1 Build the case and install the RPi and BrickPi inside it (see Figure 10.61). Dexter Industries has a tutorial showing how to do this: http://www.dexterindustries.com/BrickPi/getting-started/basics/.

FIGURE 10.61 The RPi and BrickPi fit inside the acrylic case.

STEP 2 Attach the case to the robot using zip ties, metal hardware, or Mindstorms connector pegs. I used a laser-cut piece of wood with the RPi case screwed into it, shown in Figure 10.62. I then attached the wooden plate, which features Mindstorms hole spacing, onto the robot using 3M peg with cross hole elements.

FIGURE 10.62 Attach the case to the robot.

STEP 3 Install the OS, which as mentioned, amounts to putting the OS on an SD card, shown in Figure 10.63. You must use Dexter Industries' modified Raspbian image or the BrickPi won't work. You can learn how to prep the SD card on Dexter's site: http://www.dexterindustries.com/BrickPi/getting-started/pi-prep/.

FIGURE 10.63 Load the operating system onto the SD card.

STEP 4 Plug the Mindstorms wires into the BrickPi, including the ultrasonic sensor wire into port S2, as well as the motors into MB and MC (see Figure 10.64).

FIGURE 10.64 Plug in the Mindstorms wires.

STEP 5 Wire up the altimeter as shown in Figure 10.65. Plug the Vin pin into 3.3V (pin 1) on the BrickPi. Plug the altimeter pin marked SDA into pin 3 on the BrickPi, and the SCL pin into pin 5 on the BrickPi. Finally, Connect GND to GND (pin 6) to ground the sensor.

FIGURE 10.65 Wire up the altimeter.

STEP 6 Add the battery pack, shown in Figure 10.66. I mounted mine on a wooden plate with Technic-spacing holes, making it easier to mate LEGO and non-LEGO parts. The pack just plugs into the RPI's power plug.

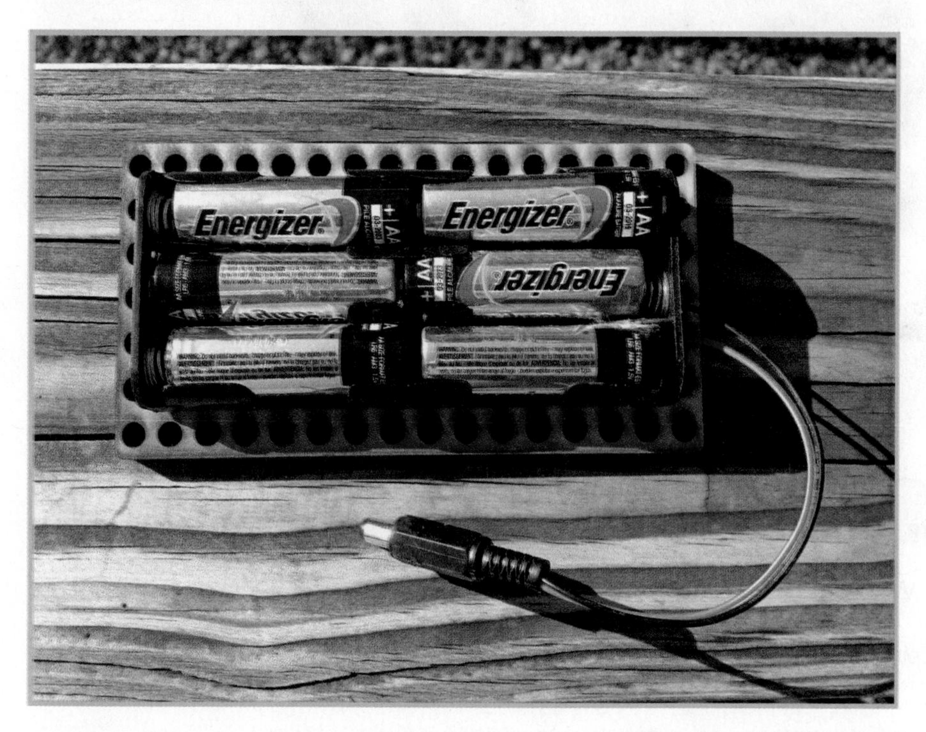

FIGURE 10.66 Add the battery pack.

STEP 7 Program the Climber. Place the following program on the Raspberry Pi and launch it to begin operation. Figure 10.67 shows the finished robot in operation!

#thanks to Dexter Industries and ClAduck for their code examples.

```
    #!/usr/bin/env python

from BrickPi import *    #import BrickPi.py file to use BrickPi operations

BrickPiSetup()  # setup the serial port for communication

BrickPi.SensorType[PORT_1] = TYPE_SENSOR_ULTRASONIC_CONT   #Set the type of
sensor at PORT_1

BrickPi.MotorEnable[PORT_A] = 1 #Enable the Motor A
BrickPi.MotorEnable[PORT_B] = 1 #Enable the Motor B
```

```
BrickPiSetupSensors()   #Send the properties of sensors to BrickPi

power = 0

while True:

#    print "Running Forward"

    BrickPi.MotorSpeed[PORT_A] = 200
    BrickPi.MotorSpeed[PORT_B] = -200


    ot = time.time()
    while(time.time() - ot < 3):    #running while loop for 3 seconds
        result = BrickPiUpdateValues()       # Ask BrickPi to update values
for sensors/motors
        time.sleep(.1)  #pause a moment
        from smbus import SMBus
        import time

# Special Chars
deg = u'\N{DEGREE SIGN}'

# I2C Constants
ADDR = 0x60
CTRL_REG1 = 0x26
PT_DATA_CFG = 0x13
bus = SMBus(1)

who_am_i = bus.read_byte_data(ADDR, 0x0C)
print hex(who_am_i)
if who_am_i != 0xc4:
    print "Device not active."
    exit(1)

# Set oversample rate to 128
setting = bus.read_byte_data(ADDR, CTRL_REG1)
newSetting = setting | 0x38
bus.write_byte_data(ADDR, CTRL_REG1, newSetting)

# Enable event flags
bus.write_byte_data(ADDR, PT_DATA_CFG, 0x07)
```

```
# Toggel One Shot
setting = bus.read_byte_data(ADDR, CTRL_REG1)
if (setting & 0x02) == 0:
    bus.write_byte_data(ADDR, CTRL_REG1, (setting | 0x02))

# Read sensor data
print "Waiting for data..."
status = bus.read_byte_data(ADDR,0x00)
while (status & 0x08) == 0:
    #print bin(status)
    status = bus.read_byte_data(ADDR,0x00)
    time.sleep(0.5)

print "Reading sensor data..."
p_data = bus.read_i2c_block_data(ADDR,0x01,3)
t_data = bus.read_i2c_block_data(ADDR,0x04,2)
status = bus.read_byte_data(ADDR,0x00)
print "status: "+bin(status)

p_msb = p_data[0]
p_csb = p_data[1]
p_lsb = p_data[2]
t_msb = t_data[0]
t_lsb = t_data[1]

pressure = (p_msb << 10) | (p_csb << 2) | (p_lsb >> 6)
p_decimal = ((p_lsb & 0x30) >> 4)/4.0

celsius = t_msb + (t_lsb >> 4)/16.0
fahrenheit = (celsius * 9)/5 + 32

print "Pressure and Temperature at "+time.strftime('%m/%d/%Y %H:%M:%S%z')
print str(pressure+p_decimal)+" Pa"
print str(celsius)+deg+"C"
print str(fahrenheit)+deg+"F"



    print "Running Backward"

    BrickPi.MotorSpeed[PORT_A] = -200
    BrickPi.MotorSpeed[PORT_B] = 200
```

```
    ot = time.time()
    while(time.time() - ot < 3):    #running while loop for 3 seconds
        BrickPiUpdateValues()       # Ask BrickPi to update values for
sensors/motors
        time.sleep(.1)              # sleep for 100 ms
```

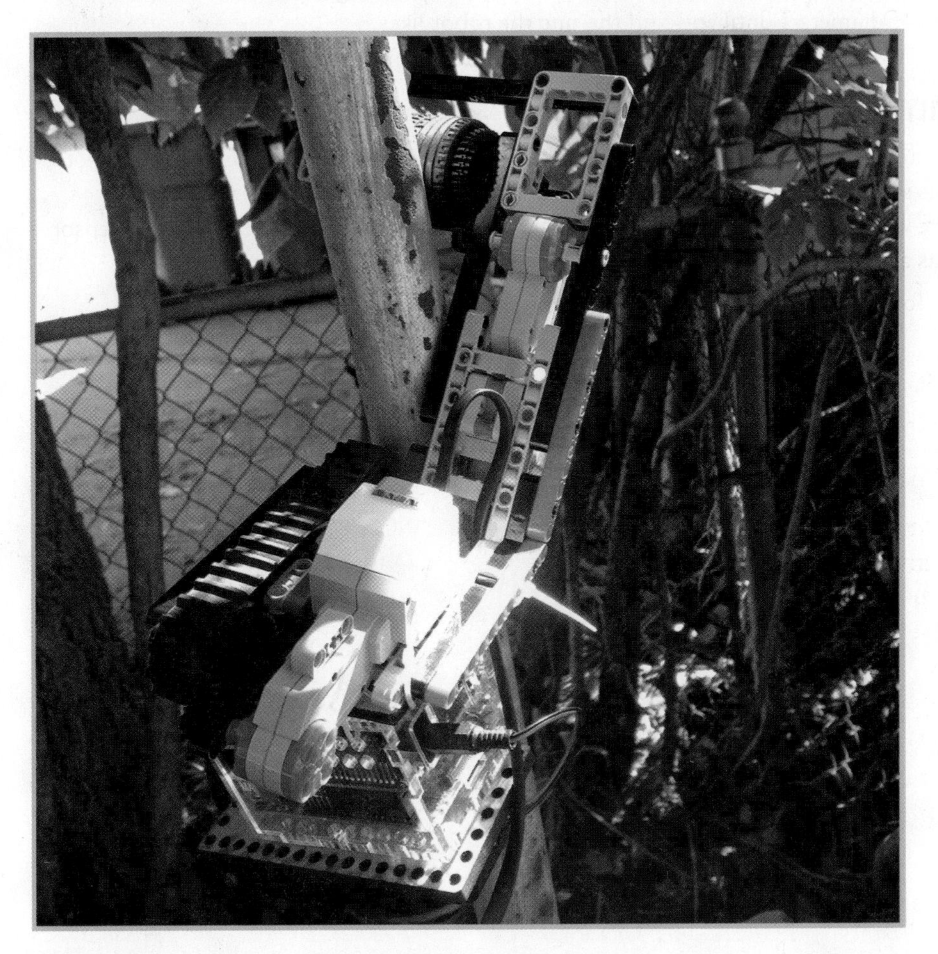

FIGURE 10.67 Program the climber and let 'er rip!

Tip: Troubleshooting

The following are some tips to ensure your climber works well:

1. The first thing to remember about the climber is that it relies on friction to work. If the flagpole in question has a smooth or wet surface you may need to either choose another pole or adapt the design.

2. The climber also relies on balance to keep the wheels on the metal. Try to keep the robot's center of gravity on the same axis as the pole or the robot will tip to one side and ultimately fall off the pole.
3. Finally, flagpole diameter does play a role in the successful operation of the robot. The climber must hang correctly to maximize the friction it brings to bear on the pole. Try different diameters until you find the one the robot likes.

Summary

This chapter was all about combining different techniques from the book: The Intelligent Brick was swapped out for a Raspberry Pi, the mounting plate I used for the BrickPi enclosure was laser-cut out of wood, and the robot made use of an NXT ultrasonic sensor as well as a third-party electronic barometer.

That's it for the book as well—thanks for reading it!

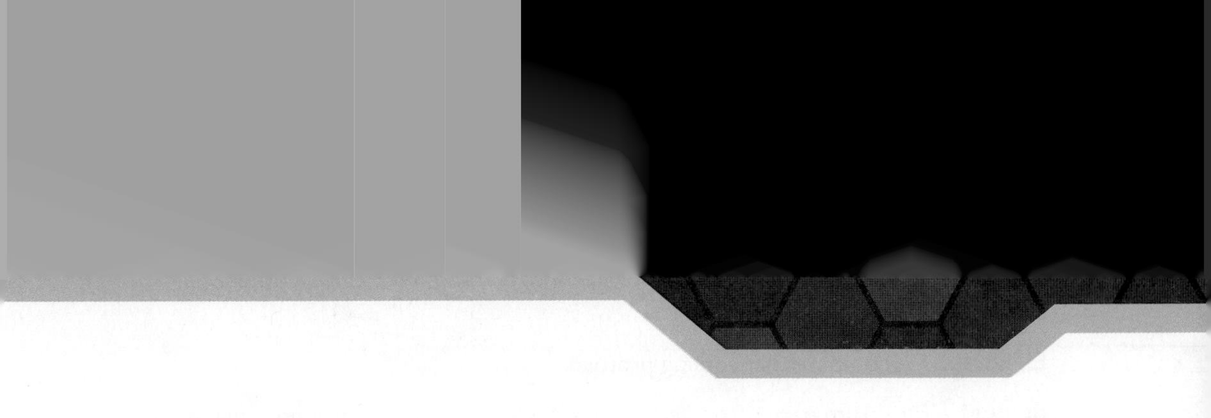

This book contains a number of terms that may be unfamiliar to a reader. Refer to this glossary any time you encounter a word or phrase you don't understand.

3D design program A program that allows the user to design an object for output on a 3D printer.

3D printer A machine capable of extruding and depositing layers of plastic to form a three-dimensional object.

accelerometer A sensor that determines its speed and acceleration, and returns that value to a microcontroller.

actuator Motors and other mechanical assemblies triggered by electricity, pneumatics, hydraulics, or some other force.

analog Data sent in a continuous wave of varying voltage, as opposed to digital, which sends data with a series of on-and-off signals.

Arduino Microcontroller board designed for artists and beginners.

BeagleBone Black Microcomputer about the size of a credit card but capable of behaving like a fully-fledged Linux PC.

Bionicle eye A largely cosmetic part (P/N 41669) found in the EV3 set.

BrickPi An interface board between a Raspberry Pi microcomputer and Mindstorms' proprietary motors and sensors.

bushing Small sleeves designed to secure the end of an axle. The EV3 set comes with red bushings and yellow half-thickness bushings.

cape In the BeagleBone Black context, it's an add-on board that adds sensors and other capabilities to the base unit.

Cartesian coordinates The system that describes a specific point in a three-dimensional space. This is represented as X (left and right), Y (forward and backward), and Z (up and down).

chassis brick Also known as a beam frame, the chassis brick consists of a rectangular beam in the shape of a hollow rectangle and is good for making a vehicle chassis and for strengthening a complicated structure.

computer numerically controlled (CNC) tools Computer-controlled power tools that precisely follow paths as directed by a computer program.

dongle A small device that can be plugged into a computer to give it additional capabilities.

Ethernet A computer networking protocol.

ground The return path of an electric circuit. On a battery the ground is marked with a - (minus sign). Ground is often abbreviated GND in electronic parlance.

HDMI A standard for connecting high-definition televisions to peripherals.

hub wheel An electrically powered wheel with the motor contained inside the wheel itself.

infrared (IR) light A bandwidth of light outside the visible range for humans; IR light is often modulated to send small amounts of data—for instance, the "off" signal for a TV.

Intelligent Brick The control brick that ships with the Mindstorms EV3 set.

iteration Each successive version of an existing project is an iteration.

laser cutter Also known as a laser etcher, a laser cutter burns through thin materials such as cardboard, MDF, and particle board.

LeJos An alternate operating system that can be installed on Mindstorms Intelligent Bricks.

linear actuator A motor that extends and retracts a rod rather than turning an axle.

mechanum wheel A wheel with smaller wheels along the rim, allowing a robot to move sideways as well as forward and backward. Also known as omni wheels.

micro USB The smallest standard size of USB cable.

microcomputer A circuit board computer capable of running Linux, connecting to the Internet, and running programs.

microcontroller A simplistic computer, capable of taking input from sensors and activating motors and lights.

Molex plug A plastic socket system for securely connecting two electronic components.

multiplexer (mux) A device capable of taking input from multiple sources and sending it to a controller via a single data connection.

open-source hardware and software Electronics projects where the code and electronic designs are shared freely, and anyone is free to modify or recreate it.

parametric In the 3D printing world, this term refers to interactive designs that can be modified by the end user. For instance, changing the diameter of a wheel.

plotter A drawing machine that uses a pen to draw shapes on a sheet of paper.

pneumatic valve A valve that controls a pressurized air system.

proximity sensor A sensor that detects a nearby object, often by sending out a pulse of infrared light and sensing the reflection.

Raspberry Pi A microcomputer capable of running Linux, connecting to the Internet, and other tasks computers perform.

reflash To reprogram the firmware on an EV3 Intelligent Brick.

relay An electromechanical switch capable of controlling high-voltage circuits.

RGB LED A light-emitting module consisting of three elements, one each of red, blue, and green. By lighting one or more of these elements, a large variety of colors can be created.

SCL (serial clock) One of the two wires of a serial data connection.

SDA (serial data) One of the two wires of a serial data connection.

servo A motor equipped with a gearbox and encoder, enabling precision control of how far the motor's shaft turns.

sketch Arduino parlance for the program that controls the Arduino's pins and contains instructions for the microcontroller's processor.

STL A common 3D printing file format.

thermal infrared sensor A sensor that detects movement by scanning changes in infrared light in its proximity.

torque The twisting force exhibited by a rotating object.

ultrasonic sensor A sensor that detects obstructions and measures distances by transmitting a beam of inaudible sound and then listening for an echo.

worm drive A spiral-shaped gear assembly featuring a low-speed but high torque.

Index

Symbols

A

B

C

D

R

S